SPACES OF CRISIS AND CRITIQUE

ALSO AVAILABLE FROM BLOOMSBURY

Topophobia, Dylan Trigg
Foucault and Power, Marcelo Hoffman
Bare Architecture, Chris L. Smith
Space After Deleuze, Arun Saldanha
An Epistemology of Noise, Cecile Malaspina
Metanoia, Armen Avanessian and Anke Hennig

SPACES OF CRISIS AND CRITIQUE:

Heterotopias Beyond Foucault

EDITED BY
DAVID HANCOCK,
ANTHONY FARAMELLI, AND
ROBERT G. WHITE

BLOOMSBURY ACADEMIC
LONDON • NEW YORK • OXFORD • NEW DELHI • SYDNEY

BLOOMSBURY ACADEMIC
Bloomsbury Publishing Plc
50 Bedford Square, London, WC1B 3DP, UK
1385 Broadway, New York, NY 10018, USA

BLOOMSBURY, BLOOMSBURY ACADEMIC and the Diana logo are trademarks of Bloomsbury Publishing Plc

First published in Great Britain 2018

Cover design by Irene Martinez Costa
Cover image © Mr Doomits / Alamy Stock Photo

A catalogue record for this book is available from the British Library.

A catalog record for this book is available from the Library of Congress.

ISBN: HB: 978-1-3500-2112-9
ePDF: 978-1-3500-2113-6
eBook: 978-1-3500-2111-2

Typeset by Deanta Global Publishing Services, Chennai, India

To find out more about our authors and books visit and sign up for our newsletters.

CONTENTS

ACKNOWLEDGEMENTS

The genesis of this book dates back to January 2015, the date of the inaugural London Graduate School conference at Kingston University. Our call for papers referenced 'Of Other Spaces', and while contributions engaged with the spatial from a number of different approaches, it was Foucault's provocation to which we returned, seeing it as ripe for our contemporary moment. While the conference provided the background to this collection, the essays within it stand alone as remarkable engagements with Foucault's call to think through contemporary spaces of crisis and critique. So firstly, we would like to thank our contributors for their hard work in producing a collection of challenging and strikingly original essays. Our thanks also go to the London Graduate School for their support and encouragement, particularly Martin McQuillan and Simon Morgan Wortham. We would also like to thank Eleni Ikoniadou.

Finally, we extend our gratitude to everyone at Bloomsbury for their support in overseeing this project from proposal to publication. Particular thanks go to Liza Thompson and Frankie Mace. We would also like to thank Monica Sukumar at Deanta for her support and guidance.

Introduction

'Make America Great Again!', 'Take Back Control!' The slogans that haunted the political ruins of the Anglophone world in 2016 sounded like those of a monstrous distortion of Benjamin's Angel of History. The hated, elitist winds of progress blowing into its wings; its face turned away from the future, contemplating not catastrophe, but an inexistent 'idyllic' past of whiteness, disequilibria of power, and rigid hierarchies. Yet despite this retro-nationalism, with its fetishization of Colonial might, 2016 saw the latest in a series of laterally connected spatial shocks. The seismic eruptions of Brexit followed by the election of Trump confounded psephologists, the media, and the political system. A post-mortem was conducted as to the causes, while Western Europe reinforced its foundations anticipating aftershocks and further tremors. Despite the evidence of topological trauma, a frantic search for causes looked to the historical – those 'left-behind' by globalization – as if the lateral irruptions of fascism and resistance, from the Middle East to Europe and the United States, could be reduced to a notion of those at the back of a linear 'historical queue'. Doreen Massey had warned us of the discursive and geopolitical bias that subordinates space – particularly in the contemporary logic of neoliberal capitalist globalization – and flattens difference and heterogeneity into points on a line:

> They are not recognised as coeval others. They are merely at an earlier stage in the one and only narrative it is possible to tell. That cosmology of 'only one narrative' obliterates the multiplicities, the contemporaneous heterogeneities of space. It reduces simultaneous coexistence to a place in the historical queue. (Massey 2005: 5)

This passage, while predating the financial crisis of 2008, articulates the lack of capacity contemporary political thinking has for understanding the networked spatial crises of the contemporary world. In order to be able to adequately respond to our present epoch, this book aims to demonstrate how 'thinking the spatial in a particular way can shake up the manner in which certain political questions are formulated' (ibid.: 9). A spatial reformulation of the political is now, we argue, critical.

The neoliberal epoch is permeated by an evolving and growing crisis. From financial meltdown and exploding inequality, governmental collapse and the dismemberment of the old party–political alliances, to civil war, the splintering of international alliances, and environmental cataclysm, all marked by the rise of global fascisms. We are living through a crisis that is shattering the basis of modernity and the assumptions that undergird it. The social organization of space no longer seems capable of containing the libidinal energies within it. Crises no longer erupt, instead they permeate everything. The onset of secular stagnation, surrounded by the generalized neoliberal crisis, is collapsing the grounds of modernity itself: progress is now doubted. Despite this, a temporal perception is homogenous insofar as there seems to be no hope of any form of life beyond capitalism, only a fast disappearing future. This homogenization of time has had the effect of emptying thought of its critical capacity.

This book is an attempt to reinvigorate spatial critiques of our present moment that think outside of the temporal ordering implicit in modernity and crucial in neoliberalism. We see the solution to this impasse as a stepping towards spatial modes of examination that create heterogeneous spaces from which we can investigate and be in the world. We recognize at least two previous important spatial interventions in critical thinking. The first, in 1967, was ushered in when, at an architectural conference in Paris, Michel Foucault gave a keynote address titled *Des Espace Autres* (*Of Other Spaces*). In this short, provocative and, at points, rambling paper

Foucault asserted that while the nineteenth century was a period obsessed with history, our present epoch will 'perhaps be above all the epoch of space' (1986: 22). Far from privileging theories of space over theories of time, this approach sought to spatialize our thinking of time to better understand the organization of networks of power and control that formed throughout the twentieth century. In doing so Foucault laid bare a line of inquiry that haunts his entire oeuvre.

Arguing that we do not inhabit 'empty and homogenous space' (ibid.: 23), Foucault, foreshadowing the work done in network theory by thinkers like Castells and Galloway, claimed space is intrinsically heterogeneous and networked by sets of relations (ibid.: 24). Among these spaces, Foucault was particularly interested in what he terms 'heterotopias', these being

> certain [spaces] that have the curious property of being in relation with all the other sites, but in such a way as to suspect, neutralize, or invent the set of relations that they happen to designate, mirror, or reflect. These spaces, as it were, which are linked with all the others, which however contradict all the other sites, are of two main types. (ibid.)

Apart from the introduction to *The Order of Things* Foucault never again engaged with heterotopias in an explicit way. However, this paper gave birth to numerous attempts to examine space differently than previously attempted and opened discourses of space to philosophy, postcolonial studies, critical theory, and critical geography. The concept of the heterotopia was engaged with in a sustained way during the early 1990s by the geographer Edward Soja through his concept of 'Thirdspace'. Following Henri Lefebvre's triad of spatiality, that is, conceived, perceived, and lived space, Soja posits Thirdspace as a critical opening. Thirdspace is imagined as critique but also as a move beyond the dominant dualism of spatial thinking conceptualized as the concrete materiality and social praxis of 'firstspace' and the discursive mediated representations of 'secondspace'. Soja conceives Thirdspace as providing 'an alternative "postmodern geography" of political choice and radical openness' (1996: 63), allowing for new emancipatory and radical modes of spatial thinking. Soja proposed Thirdspace as radically open sites that critically reflect back the dominant spatial ordering. Thirdspace is

> where issues of race, class, and gender can be addressed simultaneously without privileging one over the other; where one can be Marxist and post-Marxist, materialist and idealist, structuralist and humanist, disciplined and transdisciplinary at the same time. (1996: 5)

'Crisis' and 'critique' share the Greek root *krino*, meaning 'to decide or separate'. Heterogeneity and criticality are thus always already present in both terms. The title of this book, *Spaces of Crisis and Critique*, is both an attempt to recognize the spatial nature of crisis and reawaken the criticality of disruptive spatial thinking that synthesizes the two interrelated spatial approaches of Foucault and Soja.

Foucault conceptualized heterotopias as largely being either spaces of crisis or of deviation, and problematically wrote that heterotopian spaces of crisis were predominately found in 'primitive societies'. These were 'privileged or sacred or forbidden' spaces for 'individuals who are, in relation to society and to the human environment in which they live, in a state of crisis: adolescents, menstruating women, pregnant women, the elderly, etc.' (1986: 24). The space of crisis is created when the individual goes through something that is perceived as a crisis by society. The locus of the crisis is the individual, created by a state of abjection. The crisis individual leaves society so as not to infect it and creates a heterotopia by doing so. Foucault goes on to note that these spaces have largely disappeared from modern society, having been replaced with spaces of deviation such as care homes for the elderly, psychiatric hospitals, and prisons (ibid.: 25).

Crisis is a term that haunts Soja's work. In fact, it seems impossible to conceive of Thirdspace without acknowledging a corresponding crisis. Soja helpfully gives several concrete examples of crises, such as the disciplinary crisis in geography and the political and civil rights crisis of racism in the United States. For Soja, these multiple crises can be grouped together in a general 'crisis of post-modernity' (1996: 19–20). For Soja, this is a crisis defined by an era where

> [cities] and all our lived spaces have been shifting from a period of crisis-generated restructuring to the onset of a new era of restructuring-generated crisis, a crisis deeply imbricated in the post-modernization of the contemporary world. As with all times of crisis, there are both new dangers and new opportunities unleashed by the multiplicity of confusing and often brutal events that have been shaking the world since 1989, from the fall of the Berlin Wall and the collapse of the Soviet Union, to the repercussions of what happened in Tiananmen Square [*sic*] and in the Los Angeles uprising of 1992, to current developments in Bosnia and in the Republican Revolution in the United States. The ultimate goal of Thirdspace and the continued journeys to the Postmetropolis is to contribute to the progressive resolution of at least some of the problems associated with this contemporary restructuring-generated crisis. (ibid.: 23)

By acknowledging the accumulated history of the twentieth century, but locating the crisis of post-modernity as beginning with the fall of European communism in 1989, Soja is explicitly arguing that this is an ideological crisis. This creates a political imperative to think spatially in order to create counter-sites of resistance to neoliberal capitalism. The generalized crisis of neoliberalism constitutes society itself rather than being a contingent development. In this context Soja's positioning of the crisis of post-modernity as ideological appears as self-evident. It is our contention that understanding heterotopias as networked spaces means that crisis cannot be reduced to an individual experience but needs to be formulated as a collective encounter with the difficulties and dangers that arise in political, economic, and cultural spaces. If we are to accept Soja's and Foucault's argument that heterotopias or Thirdspaces are primarily counter-sites, then the politics that follows becomes one of critical resistance.

Foucault tells us that this form of resistant politics is both spatially organized and critical when he conceives of heterotopias as counter-sites where, 'the real sites, all the other real sites that can be found within the culture, are simultaneously represented, contested, and inverted' (1986: 24). It is from and through these spaces that a culture can be critically examined as part of a response to a crisis. This is the same critical function which Soja is interested in when he proposed Thirdspace as a process of 'thirding-as-Othering'. Soja describes this as an attempt to

> [open] up our spatial imaginaries to ways of thinking and acting politically that respond to all binarisms, to any attempt to confine thought and political action to only two alternatives, by interjecting an-Other set of choices. In this critical thirding, the original binary choice is not dismissed entirely but is subjected to a creative process of restructuring that draws selectively and strategically from the two opposing categories to open new alternatives. (ibid.)

The opposing, pre-existing categories are the analysis of space through the lens of history (temporality) and through the social (ibid.). The introduction of an 'Other' space opens up the previously privileged categories of theoretical and political analysis to new and creative forms of critique that would have the ability to guide a new and potentially emancipatory praxis through allowing, as Soja puts it, 'the translation of knowledge into action in a conscious – and consciously spatial – effort to improve the world in some significant way' (ibid.: 22).

Foucault counterpoises the concept of heterotopia with the more traditional concept of the utopia. The other-place of heterotopia is distinct to the no-place of utopia, which Foucault calls 'fundamentally unreal spaces'. Utopias do *not* exist, but the heterotopia *does*. Foucault calls them 'places that do exist and that are formed in the very founding of society'. The utopia is a political imagining of a world that is not but could be, utopias 'present society itself in a perfected form'. Utopias are thought-experiments, fantasies, and fictions that as well as performing a form of critique of the present also provide a political function. Utopia drives a politics of the present towards a future meaning that they are crucial to a temporal politics.

The age of austerity has reawakened an interest in utopian thought through a desire to imagine a non-neoliberal world. These thoughts quickly tread an old path that leads nowhere. Utopian ideas drive political projects forward towards a promised land and as such they operate as a mechanism of political excess. Utopias provide a justificatory apparatus of sacrifice. So goes the critique of Soviet communism or Chinese Maoism. This critique is correct. Both systems produced great feats but at great human cost. We may postulate that without the utopian idea, without the promised future, the human misery accepted in their name may not have taken place. There is a danger in the desire to recapture the utopian impulse. Utopia as a political technology isn't owned by a particular point on an ideological spectrum. We should not forget that Hayek, the philosophical lodestar of neoliberalism, recognized early on that the overthrow of the post-war Keynesian consensus would require something more than an economic argument (Hayek: 1967). It needed a myth; it needed a utopia. The price was worth paying for the deregulated paradise to come. The dismal politics of austerity follows this creed. The delusions of Brexit and Trumpism are similarly utopian; they exist not in the world but in a temporal fantasy which will demand its price. Utopia then is a double-edged sword. It presents a competition between visions of the future. It is, in other words, a marketing exercise.

In this sense heterotopia is distinct. Spatial, not temporal, the heterotopia does not imagine a future (or a past idyll) but requires us to look at what is. Heterotopias are real spaces that exist 'outside of all places, even though it may be possible to indicate their location in reality' (1986: 24). Using the analogy of a mirror, Foucault goes on to describe heterotopias as spaces that are real and virtual at the same time, insofar as they reflect an image that allows an individual or society to see themselves in a place where they are not. Heterotopias are 'absolutely real, connected [networked] with all the space that surrounds it, and absolutely unreal, since in order to be perceived it has to pass through this virtual point which is over there' (ibid.).

The spatial aspect of heterotopias allows for an entirely different form of thought – it allows us to think *Otherly*. Whereas utopian thought relies on teleological notions of progress, thinking spatially is grounded in a materialist phenomenology; it is a perceptual form of critical thought that occupies space at the same time as it constructs space. It is a domain of thought established in a borderline environment that constructs an imago of the world as an organization of knowledge. This means that spatial thinking is acutely aware of borders and limits, but also attentive to breaks or potential openings. As such this is a form of thinking that can perceive possibilities that do not fall back on myths of progress. Following from this, heterotopian thought, unlike utopia, is not necessarily political. Heterotopias are not intentional, and they do not need to be part of a project. Heterotopias exist as different spaces which reflect and contest reality but often without conscious construction. The consideration of heterotopias is therefore wider than simply political, they can be aesthetic and bodily as well.

Indeed, while this book stresses the need to think the spatial anew through heterotopia, it also calls for a rehabilitation of spatial thinking as an aesthetic as well as political process. That is to say, the 'spatial turn' from Foucault through Edward Soja, Doreen Massey, and David Harvey has thus far been dominated by the perspective of critical geography; as such any articulation of the spatial is through a geometrical (and Newtonian) paradigm, as Smith and Katz (1993) have recognized. They argue that this has led to a tendency to employ spatial metaphors while borrowing uncritically from the language of absolute space. Arguably, this overemphasis on cartographic, absolute space as a source leaves any 'new' spatial thinking in a cul-de-sac. Further, this dominance of critical geography in answering Foucault's call in *Of Other Spaces* has been at the expense of spatial aesthetics. This is despite this piece *implicitly* referencing the aesthetic (and the corporeal) throughout (the garden, the Persian rug, old age) and *explicitly* referencing Bachelard's *The Poetics of Space*, a work on the aesthetics of the infinitesimal and vast spaces of the material and mental. It is therefore, we believe, vital to rearticulate the potentiality of Foucault's spaces of critique for thinking the aesthetic *through* the political. This book therefore considers the heterotopia from the aesthetic perspective and places these chapters at the front of the book so as to frame the later discussion of politics, because while heterotopias may not necessarily be themselves political in intent, in their existence as heterotopian spaces they act as modes of critique and in this sense they embody a challenge to reality.

The book opens with Fred Botting's chapter 'Literature-Outside-Space: Foucault, Sade, and Tales of Terror' where Botting investigates the 'heterotopical implications' of literature's sub-critical interdictional space of the 'Tales of Terror'. Botting's chapter opens up Foucault's provocative work to the monstrous and abysmal heterotopic spaces in literature. These spaces claim a non-human sovereignty that slips away from a modernist ethics of rationality and sensibility.

The Order of Things begins with the shattering laughter provoked by a literary mode associated with the collapse of every discursive formation associated with 'Man': it releases a monstrosity that – heterotopically – defies as much as constitutes any ordering mechanism of identities, relations or classifications. A literature which is not yet literature murmurs throughout Foucault's text, beneath, within, and outside the polarizations and discontinuities assembling as the life, labour, and language of human science and calling up an ill-formed but formative sub-critical space of doubling, repetition, return; differences reduplicating difference, unthought troubling thought, and a strange insistence of some kind of outside.

If the elaboration of literary space and effects subtends the entirety of *The Order of Things*, its heterotopical implications are found in the essay on 'Other Spaces'. But literary virtuality (reduplicating the virtual nature of any linguistic encoding) attenuates another non-place, often invisible in patterns of literature's historical and generic emergence and monstrosity: un-spaces of hyperbole and limit echo the anguish of transgression, zones of both deficiency and excess open abyssal non-relations ('Preface to Transgression'). Indeed, in the movements of language beyond frames of mimesis and representation, an enunciation, of nothing, from nowhere, signals a writing 'to infinity' composed of mirror-dense effects, transparent affects, and a proliferation of affirmative terrors.

Literature finds voice in writers who have most effectively plumbed its abyssal surfaces: Sade, Blanchot, and Bataille, of course, predominate ('Language to Infinity'). There are other writers, too, known to Sade, who offer exemplary instances of literature's sub-critical interdictional space: generally unnamed, sometimes misnamed, they are grouped darkly under the genre that defined them: the 'tales of terror'. Their writing – detritial prelude to the literature of progressive modern rationality, sublime sensibility and ethical subjectivity – prefaces a finitude in which something sacred flickers and something sovereign squirms and slips away briefly disclosing a space of disarmingly affirmative, neutral, and negative potential.

Julian Reid expands upon the aesthetic implications of Foucault's work in 'The Living Space of the Image'. Moving away from Botting's investigation of non-living, un-dead space, Reid looks at the spaces that images occupy in relation to the politics of life and the world. Foucault is read through a phenomenological understanding of 'the world' as the space that we live in, but understood as a living image. Reid asks: What sort of space do images occupy and how does that space relate to the politics of life and world? The world, as Merleau-Ponty maintained, is what we live. But in living it we live off the life of images. For images are also, in an uncanny way, of this world; necessary resources for us in our quests to penetrate its truth, while not to be confused, we are told, with the world as such. Instead they occupy a 'stage' out there 'in front of the world' guiding us to the real while forever preventing us from seeing the real as such. We encounter a similar idea, of the errant spatiality of the image, in the philosophy of Guy Debord when he argues that images are not somehow 'out in front' of the real but 'exist above it' while simultaneously imposing themselves upon it. The life of the real, we must suppose, exists either beyond the stage of the imaginary or below it. How can we tell the difference between what is real and what is image when the image is both a necessary resource for guiding us as to the nature of the real as well as being that which mystifies it? What are we to make of the conflicting ways of construing the spatiality of images in the philosophy of imagination? Reid's powerful contribution answers these questions.

Sheena Culley's chapter, 'Inside Comfort: The Interior and the Immune System', contrasts Reid's examination of the living world to the enclosed space of the living body. Numerous attempts to define comfort have left us with a negative definition of the idea only in relation to discomfort. In addition, comfort has been defined by John Crowley as 'a self-conscious satisfaction between one's body and its immediate physical environment'. This definition sets up a relationship between dualistic terms; body and environment, which also imply an interior and an exterior. In cultural and historical scholarship, comfort is posited as that which privileges interiority, from Walter Benjamin's observations of nineteenth-century dwelling to John Gloag's notion of

Victorian comfort which became a matter of 'exclusive privacy'. Yet, matters of space which relate to the body are under-recognized in existing literature on comfort. In this chapter, Culley argues that ideas relating to immunity and immune systems have also enforced the primacy of the inside in the idea of corporeal comfort. Drawing on historical and contemporary ideas of immunity in Ed Cohen's and Peter Sloterdijk's scholarship, Culley asks if a different definition of comfort can be sought which does not see the outside in opposition to the inside.

Building on Culley's critical approach to the inside/outside dualism of the enclosed space of the body, Robert G. White's chapter, 'Spacing the Interior: The Carceral Body as Heterotopia in Contemporary Palestinian Cinema', examines the biopolitics of discipline that necessitates the separation and distribution of individual bodies in space in order to render them docile. In *Of Other Spaces* (1967), Foucault defines heterotopias of deviation as 'those in which individuals whose behaviour is deviant in relation to the required mean or norm are placed'. White reads Foucault's spatial ordering of (docile) bodies with Jean-Luc Nancy's (2008) 'spacing [as] tension of place where bodies are not in space but space in bodies'. The world of bodies, Nancy writes, 'is a world where bodies initially articulate space'. The body becomes an enclosure, a heterogeneous space of disciplinary monotony. From here, this chapter explores the Palestinian 'interior' cinematic body in crisis, a body *becoming* a carceral space. Carceral bodies articulate biopolitical space, becoming heterotopic sites in a network; not bodies *in* space but *as* space. This concept is explored through a close reading of the films of Kamal Aljafari and Elia Suleiman. The chapter concludes with an analysis of spaces of resistance in these directors' work. Foucault's heterotopia is read through Agamben, arguing that a topological reading of heterotopia opens up new paths to thinking resistance.

David Hancock's chapter, 'The Politics of the Hidden Space: Georges Bataille and non-knowledge in the era of transparency', looks to conceal the enclosed space of bodies in order to articulate a politics of hiding. Hancock begins with the proposition that we are in an era of transparency where openness and visibility are taken as absolute goods and a cornerstone of the democratic system and are crucial for neoliberal governance – while secrecy, invisibility, and hiddenness are received with suspicion. Building on these ideas, Hancock asks the question of the politics of these hidden and no-longer hidden spaces through a re-examination of the sociology of the secret society and epistemology of non-knowledge.

Hancock's contribution will focus the secret society through the sociology of Georges Simmel, Roger Callois, and Georges Bataille. What is revealed through these authors is the apparently intractable divide between transparency and the political. Transparency gives the appearance of legitimacy but neutralizes politics. Secrecy allows prudent politics but also creates space for mendacity. Secrecy and the ability to remain hidden, as well as giving protection, allow the production of new forms of politics and modes of being to take shape.

The second part of this chapter looks at this apparent aporia through a more detailed consideration of Bataille's reflections on non-knowledge. Following his thoughts on secrecy and the establishment of the secret society Acephale in the 1930s, Bataille turned away from the politics of the hidden to the not-known. Bataille explodes the binary between what is known and what is not known and between having and not having knowledge by describing the experience of non-knowledge itself as a fundamental experience of being. The system of knowledge, its production, and its agnatological destruction as well as the production of the visible and the invisible are always already wrapped up in project and thus never move beyond the binary discourse. Through Bataille's non-project, the invisible, the secret, and not-known become a limit experience and not merely part of discursive practice. Non-knowledge is not a sociological or scientific problem to overcome or strategically create; it is a profoundly unsettling experience of existence. Secret and hidden spaces thus offer a richer form of life beyond the radical transparency of the neoliberal.

Reversing Hancock's politics, Anthony Faramelli's chapter, 'Mirrors and Masks: The Political Space of Zapatismo', looks at the politics of bringing bodies into an open space, paradoxically through a technology of hiding. For Foucault the mirror best illustrates heterotopic other space by opening up a virtual zone that is grounded in reality where people see themselves in a space where they are not, where they are absent. The mirror presents a virtual inverted realm of possibilities, allowing people to see how they could be somewhere else (ibid.: 4).

The contemporary writings on the heterotopic space of mirrors have found a home in the political philosophy of the Zapatista insurgents, zapatismo. In the Zapatista writings, reflections can be bounced off of multiple mirrors in order for politicians to hide corruption, but mirrors can also be used to reveal the ugly 'true' face of neoliberalism and present a virtual, utopic Mexico to the world. Most strikingly, the Zapatistas also write that their iconic pasamontaña (ski mask) is a mirror that opens up the expansive space of 'Mexico from below' where different resistants can find each other and connect to the global insurrection against neoliberalism. In this way zapatismo's heterotopic mirrors have reinvigorated the sense of crisis, or transformative potential, that Foucault noted was the foundation of heterotopic spaces.

This contribution elaborates on zapatismo's heterotopic spaces, arguing that zapatismo's spatial politics has facilitated a wholly original approach to political resistance. Faramelli goes on to argue that, as a political assemblage, the pasamontaña has opened up a heterotopic space where different singularities are united by a common desire for the world of Mexico, a world not dominated by neoliberalism's reduction of life to the principles of precariousness and control.

Claire Colebrook's chapter, 'In the Beginning All the World was America', closes the book by thinking through the opposing idea of Mexico as an apocalyptic space of darkness and despair presented by Donald Trump and the United States Republican Party. Taking the 2016 Republican National Convention where the party officially nominated Donald Trump to be their presidential candidate as her starting point, Colebrook looks at the Republican's quasi-religious Manichean conception of an 'America of light' locked in an eternal struggle against a 'Mexico of Darkness'. Arguing that both images present us with abysmal spaces of light and dark, Colebrook seeks to fully unpack the heterotopic spaces of America and Mexico in order to see it beyond reductive ideological representations. In essence, Colebrook wants to think through the meaning of America as an idea, *and* as an idea of light and self-making beyond which there is darkness in order to answer the question: What would it mean to take this claim non-ideologically or literally?

This question leads Colebrook through a broad philosophic survey, thinking alongside figures like Agamben, Arendt, Heidegger, Derrida, Deleuze, and Guattari to argue that politics that locate political project's beginnings necessarily fall into apocalyptic imagery that calls for fascistic intervention to adjust reality as such. Colebrook offers a politics of immanence as the only viable alternative to the death-drive of this form of liberal fascism. This final chapter forecloses the possibilities of a politics concerned with the living world and, returning to Botting's opening chapter, works as a political intervention that thinks through the sovereignty of the world as a non-human and abysmal space.

The wide range of disciplinary positions adopted by the chapters in this book reflects that transdisciplinary nature of spatial critique. It is our aim to present these chapters as an invitation to further spatial interventions in politics, art, and philosophy with the hope of breaking out of the foreclosed spaces of control that have dominated political and philosophic thought.

References

Foucault, M. (1986), 'Of Other Spaces', *Diacritics*, 16(1): 22–7.

Hayek, F. (1967), 'The Intellectuals and Socialism', in F. Hayek (ed.), *Studies in Philosophy, Politics and Economics*, 178–94, London: Routledge and Kegan Paul.

Massey, D. (2005), *For Space*, London: SAGE.

Nancy, J. (2008), *Corpus*, New York: Fordham University Press.

Smith, N. and Katz, C. (1993), 'Grounding Metaphor: Towards a Spatialized Politics', in M. Keith and S. Pile (eds), *Place and the Politics of Identity*, 67–83, New York: Routledge.

Soja, E. (1996), *Thirdspace*, Cambridge, MA: Blackwell.

1

Literature-outside-space: Foucault, Sade, and Tales of Terror

Fred Botting

Nightmare

> A nightmare has haunted me since my childhood: I am looking at a text that I can't read, or only a tiny bit of it is decipherable. I pretend to read it, aware that I'm inventing; then suddenly the text is completely scrambled, I can no longer read anything or even invent it, my throat tightens and I wake up. (Foucault 1998: 290)

This is a striking nightmare, particularly given the status of the dreamer who recounts it. Aware of its significance in terms of a lifelong personal investment bordering on obsession, Michel Foucault describes it during an interview concerned with the writing of history. For a philosopher–historian renowned for analysing the mechanics of power/knowledge in terms of discourse, a dream in which radical and active textual unintelligibility comes to the fore is a nightmare indeed: it haunts understanding and imagination with a density that can neither be penetrated nor unravelled, delineating a limit that cannot be overcome, bypassed, or thought; it dissolves the very possibility of order and sense in written objects that assume an agency of their own and, further, engenders the reciprocal and corporeal symptoms of anxious and involuntary speechlessness in the subject incapacitated in an act of analysis failing to establish its power. No speech, no confession, no discourse, no knowledge: the nightmare seems to disclose – and close off – the possibility of any glimpse of an outside, any space of production or affirmation, any grounds for subjectivity. The nightmare also discloses something about a kind of language that underlies common experience while remaining obscure in its operations, a language 'that exists everywhere and escapes us in its very survival. It survives us by turning its looks away from us, its face inclined towards a darkness we know nothing about' (290). Pervasive yet obscure, such darkness is not a simple invitation to powers of light and enlightenment to resume their work; it does not take the form of an alluring blackness or of an ink illuminated by the whiteness of a page's spacings, but has a formless form that underlies and exceeds, allows and destroys the very possibility of distinctions between light and dark, space and inscription.

In the context of the interview, the significance of the nightmare bears upon writing as much as history: it is linked to Foucault's acknowledgement of the importance of Maurice Blanchot for any kind of literary practice/thinking. His writings, Foucault contends, 'made possible all discourse about literature' (287). The claim is based on a specific and original definition: this is not a literature already furnished with form, object, and aesthetic, not the literature that, after Romanticism, necessarily elevates human vision, interiority, and imagination in the perception and presentation of the being of the world, but a less self-aggrandizing modality of something like a language within language in which a literary energy – a kind of dark poeisis – 'constitutes the outside of every work' (287). Obscure, it is powerful also: it 'furrows all written language and leaves an empty claw mark on every text' and, like the fissure that sexuality assumes in the transgressions that define modern subjectivity (Foucault 1977b: 29), it is a defining limit, a thin dark line, 'a groove that runs like a great impulse through all literary languages' (1998: 287). As fissure and groove and particularly as an obscurely intimate and indeterminable 'outside', the space of literature is sketched amid clusters of metaphors (mirrors, doubles, voids, and murmurs) that are reiterated across Foucault's writings.

Outside

'The Thought of Outside' offers a related discussion of literature that, again, acknowledges the importance of Blanchot's writings. Foucault's relation of literature and/as outside begins with an elaboration of the devastating simplicity of a basic sentence: 'I speak'. The statement, calling up and confounding vocal anteriority in the very moment and act of speaking, discloses the absence of any supporting discourse, any grounding frame or context with which to restrain the errancy of language. Instead, it marks the site where a 'void', a 'contentless slimness', enables language to 'endlessly spread forth'. Thus outspread, language unfolds in 'pure exteriority' (1987: 10 11). Unfolding – in the definitive movement of modern literature – also involves a folding back, a 'doubling back' in which a reflexive but superficial gesture of interiorization masks a more disarming passage to the outside where language exceeds subject and discourse and escapes representation: 'the "subject" of literature (what speaks in it and what it speaks about) is less language in its positivity than the void language takes as its space when it articulates itself in the nakedness of "I speak"' (1987: 2). A 'neutral space' – a space disclosed by (and defining) Western fiction – marks a decisive break with modes of discourse set upon historical determinations of thought and truth: speaking 'runs counter' to thinking, its subject dispersed and effaced while the latter reaches for the deepest interiority and certainty of an 'I'. Eclipsed by the language it speaks, the subject disappears – via 'void', 'silence', 'rumbling', and 'negation' – in a 'pure outside where words endlessly unravel' (1987: 22). While discursive thought might try to return words to bodies, wills, meanings, objects, and interiority, fiction presses elsewhere: it 'consists not in showing the extent to which the invisibility of the visible is invisible. Thus, it bears a profound relation to space; understood in this way, space is to fiction what the negative is to reflection (whereas dialectical negation is tied to the table of time)' (1987: 23). The outside-space disclosed in and as an effect of literary movement is bound neither to thought nor to things but discloses a voiding and spacing of interstices and other sides, an elusive unravelling, an invisible, negative mirroring.

Heterotopia

The literature Foucault identifies with Blanchot is not yet literature in any critical or popular sense, even though it constitutes the space where literary works can be shelved, catalogued, circulated, and consumed, thus rendered and destroyed as objects. As a literariness without literature that evacuates classification, a sub-literature whose minority indicates a subtension and negation, it lies outside, beneath, and between form, statement, and object. The conditions it sets are neither determining nor causal, yet remain possible and virtual in the way that heterotopia operate as 'counter-sites', non-dialectical locales of relation, contestation, transgression, constitution, dissolution, movement, and transformation. The sites and spaces of Blanchot's fiction exemplify literature's relation to the outside. Neither real nor unreal (un-real, perhaps), the houses, hallways, doors, and rooms in Blanchot's writing are 'placeless places' (Foucault 1987: 24). Literature in this sense is neither positivist nor utopic. The doubling and negation of fixed location, mark of literature's dislocated and mobile locus, combines with an image of negative, mirrored reflection: the space that emerges is distinctly heterotopic in shape. In 'Of Other Spaces', indeed, the image of the mirror – itself another 'placeless place' – offers an exemplary heterotopia. Charting historical discontinuities in notions of space, Foucault's lecture distinguishes everyday spaces (and their utopic, unreal counterparts) from those real but marginalized spaces that negotiate, delineate, and consolidate uncommon states. Heterotopia are those 'other' or 'different' spaces which traverse, demarcate, bolster, and disrupt the homogeneity of all the sites and borders composing everyday material existence. Marking out transitional and heterogeneous spaces, heterotopia remains part of, yet aside from, daily life: heterotopia of crisis (adolescence and boarding school) or deviation (prison, asylum, care home) deal with different conditions often inimical to usual social forms and functions; temporal heterotopia ('heterochronies') collect (museums, libraries) or waste time (vacation villages); others compensate for reality by creating spaces of illusion or otherness (imagination, romance, adventure). Heterotopia are thus sites that stand in relation to normal zones and spheres of existence, sites distinguished for 'being in relation to all other sites, but in such a way as to suspect, neutralize, or invert the set of relations that they happen to designate, mirror or reflect'. These 'counter-sites' – virtual as well as imagined – are places where reality is 'simultaneously represented, contested and inverted' (Foucault 1986: 24). The mirror is exemplary, both real and unreal, representing, inverting, and contesting reflection in a process of reflection. Foucault notes how the mirror is 'a placeless place' in which self discovers self-image in a site where it is not, not so much as reflection, but in the space of reflection, an 'unreal, virtual space that opens up behind the surface' and as 'a sort of shadow that gives my own visibility to myself'. Though it returns oneself to oneself – I see myself 'where I am absent' – the mirror also 'exerts a sort of counteraction on the position that I occupy' (1986: 24). Here and not here; there are not there, the self that seems to be returned to itself in the closed circuit of inversion is not rediscovered in a simple movement from one to other, from other back to one, but drawn out of and away from itself at the same time.

Monster

Heterotopia pertains to the monstrosity of Foucault's childhood nightmare and seems to inaugurate a literary mode in which disturbances and doublings of writing disclose a disarming, mobile space. The recurrence of Foucault's nightmare (more than a sleep of reason or unconscious eruption) is evident in his waking and writing states, spelt out at the heart of major texts that examine the birth of 'Man', literature, and the human sciences. *The Order of Things* contends that modernity, its subject, thought, and writing is formed out of practices of knowledge that shift from notions of natural history, wealth, and representation to ideas of biology, economy, and language. Born out of the break-up of classical modes of word-use as resemblance and representation, language and literature begin to find a shape beyond systems of classification and tabulation that would hold signs and reference together on the basis of a similitude sustained by the authority of a divine Word or by the codifications in which words and things were adequated and measured against each other in terms of a regulated taxonomy of observed nature.

In the preface to *The Order of Things*, Foucault's nightmare is plotted in the full glare of day and in a manner that declares its prominence in the discursive formations of modernity. It is displayed in the text of another writer. Foucault notes the uneasy laughter accompanying his reading of Jorge Luis Borges's Chinese encyclopaedia, laughter provoked by the weirdest and most asystemic of classifications of animal life: those 'belonging to the Emperor' or 'fabulous' or 'tame' or 'sucking pigs' or 'drawn with a fine camelhair brush' or those that 'are included in the present classification' or 'from a long way off look like flies'. Categorical strangeness prompts Foucault to consider the 'powers of contagion' that are evinced in the in-distinction of real and imaginary hybrids, a 'bestiary of the imagination' in which monstrosity inserts itself in the 'empty space' and interstices of differentiation to the point that it 'transgresses the boundaries of the imagination' (Foucault 1970: xvi). In respect of monstrosity, Foucault begins where Georges Canguilhem (2008) leaves off: for the latter, there is no mixing of natural and imaginary monsters. For Foucault, the reverberating strangeness of Borges's classifications do not simply invert or deviate from norms, nature, and rigid taxonomy in a negation that serves, belatedly and structurally, to affirm subsequent ordering (monsters, he notes, do not simply mark a necessary antithesis affirming proper species and norms, that is, the 'emergence of difference' they also provide a 'ceaseless background murmur' (1970: 154–6)). Indeed, monstrosity is not a matter of real bodies or imaginative bestiaries: the 'monstrous quality' of Borges fabulous bestiary lies in the 'fact that the common ground on which such meetings are possible has been destroyed'. Borges's disarming asystematicity has undone the possibility of both separation and relation: what is at stake is not so much the 'propinquity of the things listed' but 'very site on which their propinquity would be possible' (1970: xvi). The only possible meeting place for such an array of animals lies in the 'non-place of language' in a space of un-relatedness that is neither that of nature, nor reason, nor, even, of imagination. Like the China figured in Borges's encyclopaedia, the non-place of literature provides a 'privileged *site* of *space*' (1970: xix). It is marked out as a heterotopia in the most radical of senses – a non- and counter-site that has all the horrifying effects of Foucault's nightmare: where utopia allows fables and discourse, running with language, heterotopias 'desiccate speech, stop words in their tracks, contest the very possibility of grammar at its source; they dissolve our myths and sterilize the lyricism of our sentences' (1970: xviii). Hence Foucault's laughter is coloured by an uneasiness 'related to the profound distress of those whose language has been destroyed' (1970: xviii–xix). Provoking loss of speech, radical dis-order, Borges, it seems, has disclosed the absence, the lack of grounding underlying all classification, his ataxonomia leading 'to a kind of thought without spaces, to words and categories that lack all life and place, but are rooted in a ceremonial space overburdened with complex figures, secret passages, and unexpected communications' (1970: xviii). The labyrinthine space of literature assumes a power and a darkness of its own, evocative in effect, affective in its emptiness, a space akin to those disturbing zones opened up in tales of terror, by Sade and, of course, by Borges himself: 'The Library of Babel' constitutes a 'site that is nowhere' an impossible gathering of books (Foucault 1977b: 67).

The discussion of sub-literary disturbance opening *The Order of Things* to the impossible condition that informs and un-forms all practices and systems of order highlights the 'living being' of a language at the centre and the frontiers of Western culture; it disconcerts formations underpinned by resemblance and representation, detaches from any basis in nature, thought, or order guaranteed by the predominance of a divine word:

> From now on we no longer have that primary, that absolutely initial, Word upon which the infinite movement of discourse is founded and by which it was limited; henceforth, language was to grow with no point of departure, no end, and no promise. It is the transversal of this futile yet fundamental space that the text of literature traces from day to day. (1970: 44)

Unhinged from a system of signs and similitudes, torn from an imperative to discover the final Word, it allows words to 'wander off on their own' and find a 'lonely sovereignty' (1970: 48–49).

Calibrated directly to this literary errance is another movement of and on language: as the nineteenth century reorders its world in terms of labour, life, and language, it is the latter that finds itself polished to a strict neutrality, objectivity, formalization, and transparency in the distribution of knowledge:

scientific language aims at delivering an 'unmisted mirror of non-verbal knowledge' (1970: 296). The imperative to make language subservient to representation, to the thought and ideas that dominate and determine all objects, produces a countervailing effect in which literature again asserts its force as an independent and unmasterable entity, re-emerging in a fold of its own, in writing that eschews content, context, speaker, interlocutor, and refers only to itself:

> When language was burying itself within its own density as an object and allowing itself to be traversed, through and through, by knowledge, it was also reconstituting itself elsewhere, in an independent form, difficult of access, folded back upon the enigma of its own origin and existing wholly in reference to the pure act of writing. (1970: 300)

Literature thus contests philology in a movement back to the 'untamed, imperious being of words' (1970: 300). Detached from the values of taste, pleasure, nature, and truth that ensured its circulation in the classical age, it also, in 'a ludic denial of them', breaks with the forms that ensure representational consistency. Opposed to 'all other forms of discourse', its 'precipitous existence' 'addresses itself to itself as a writing subjectivity' (1970: 301). As a writing subjectivity, surrounded by language and the literary impossibility of self-representation, the empirico-transcendental doublet that is 'Man' encounters both the finitude of time, body, and enunciation, and the singular capacity to universalize itself in the face of the limits that enable self-definition.

Terror

Literature, then, offers a powerful, repeated, sometimes terrifying murmur – like that of monstrosity – in the discontinuities of modern discursive formations. At the limits and the core of any system of ordering it marks a 'mute, unsignifying region where language can find its freedom', a literary space subtending modernity, stretching from the romance that inaugurated modern fiction – *Don Quixote* – through tales of terror and the 'plastic violence' in which, rejecting discourse, Artaud and Roussel deliver the shocks that return to cry, body, materiality, and flesh (1970: 353). From there, it moves on to the writings of Kafka, Blanchot, and Bataille where, through attraction and transgression, it again broaches the outside. At a crucial historical hiatus, however, literature comes close to the basest and most extreme forms: the writings of Sade, significantly in *The Order of Things*, distinguish an experience at the edges and limits of both classicism and modernity, a juncture where literature erupts in an orgy of reason, desire, and violence, pushing language to a point of exhaustion in search of a final word and tormenting the limits of representation with a writing untethered from any law. Sade's text offers an 'incessant primordial murmur in our culture' in which language finds its 'brute being as a thing' and is pulverized by a desire it forever exceeds (1970: 195; 118); it foregrounds 'the nakedness of desire as the lawless law of the world', 'laying desire bare in the infinite murmur of discourse' (Foucault 1986: 17). Sade's vicious and incessant murmur conjoins two modes – a language of utter transparency and a writing of infinite self-reflection – in an engagement with excess that is never fully exhausted.

The writings of Sade and the tales of terror turn on excess. A pre-eminent figure of English gothic fiction, Ann Radcliffe, whose writings were enjoyed by Sade, repeatedly cautioned, in the interests of virtue and morality, that 'all excess is vicious' (Radcliffe 1966: 20). In contrast, for Sade, as Blanchot notes, 'everything is good when it is excessive' (Blanchot 1967: 54). The site of polarization, excess is the only imperative, whether it be invoked in the interests of virtuous self-regulation or as an injunction to unrelenting, unsated transgression. The association of Sade and the tales of terror might seem morally and politically like a simple inversion, but it discloses a site of unavowable proximity: here excess not only distinguishes a limit like the otherness of visible monstrosity, a boundary which provokes horrified recoil (or demands an ecstatic swoop into the delights of a world without borders or restraints), but also signals a consciousness of a general arbitrariness, an underlying groundlessness over which any system – of nature, law, or norm – discovers itself to be both unfixed and unfounded.

In his essay 'Reflections on the Novel', Sade comments favourably on the work of two late eighteenth-century English writers of terror tales, Ann Radcliffe and Matthew Lewis, and observes that the shocks produced by their fictions echo the louder reverberations of the French Revolution. Revolutionary times require revolutionary fictions, writings that repeat rather than subdue the shocks of sublime events: to avoid the 'monotony' of the novel of misfortune, Sade urges writers to draw on more extreme resources, 'to call upon the aid of hell itself' (1966: 109). Polarization also occurs at the level of virtue: for Radcliffe, though her heroines' propriety and self-control are tested to the extreme, their fortitude and avoidance of vice and temptation is ultimately rewarded. For Sade, however, the female figure of 'virtue in distress', so prominent in the eighteenth-century novel, is too much of a temptation: the heroine's innocent, virtuous body – in *Justine* notably – is too attractive as an object for despoliation, the halo of virtue, as much as physical violation and pain, invites almost interminable violence. It is not enough to subjugate the body to a vast catalogue of vicious torments, all comforts of morality must also be stripped away in an effort that is as much political and philosophical (a mirror of categorical Kantian rationality) as it is pathological and sexual. 'Virtue', as Dolmancé declares in the monotonous sexual-educational salon exchanges that make up 'Philosophy of the Bedroom', 'is but a chimera whose worship consists exclusively in perpetual immolations' (Sade 1965: 208). Debunking illusions – like goodness, benevolence, or compassion – Sade's texts demonstrate an implacable rational impulse, every episode, atrocity, experiment, and punishment an enactment of reason at its sharpest, and always justified by extensive philosophical argumentation and in the service of Nature's destructive principle: for Sade Nature's imperative demands the selfish, criminal, murderous extraction of any pleasure from any body in any way at any time. Law, like morality, is just another chimerical prejudice blocking the path, with crime constituting the only rational imperative for action. Even, and most especially, the place of God is evacuated: it is 'empty' and occupied by an 'impostor'. In *Juliette*, the claim is reiterated: ideas are but 'representations of objects that strike us' and God – 'an idea without object' – is nothing 'other than hallucination' (Sade 1968: 304). Hallucination embraces psychology, religion, and the world of ideas, its extension disallows in advance any intimation of a truly transcendental idea or ego.

Nature informs the entire cruel rationality: 'she' is cause, law, reason, energy; she is locus of creation, destruction, and absolute indifference. She dictates a primary law of destruction that is not equivalent and opposed to its complement, creation, and is fundamentally inhuman. Nature stamps – 'casts' – humans with specific impressions and directions like 'self-preservation' and 'multiplication' but then casts them out into a world dominated by the vicissitudes of permanent destruction. Humans are 'froth', excrescences with 'no intrinsic value' while nature retains 'no control' over their monstrous development, as mere objects in the vegetable and animal worlds determined by laws that destine them to 'mechanically reproduce and destroy themselves' (Sade 1968: 765–69). Yet the cycle of natural destruction that takes Sade beyond all foundations, all grounds, and any basis in law, morality and humanity is not, however, enough. He wants more, as, apparently, does Nature. Life and death are merely aspects of a wider and destructive transmutation of matter in perpetual motion. But ever more destruction is never enough, determined by an unstoppable principle of total extinction driven by perpetual motion without a prime mover (Sade 1968: 300). And though the 'extinction of all beings would make way for the new things Nature desires' this 'perfect equilibrium' of extinction is endless and unbalanced without end, is also a perfect disequilibrium, 'a perpetual metempsychosis, a perpetual variation, a perpetual permutation embracing all things in perpetual movement' (Sade 1968: 769). It is, as Gilles Deleuze reads it, a case of two asymmetrical natures: a primary, unattainable nature of pure negation that overrides 'all laws, free from even the necessity to create, preserve or individuate' and lies 'beyond all foundation' and a 'secondary nature' of cycles of life-death, creation-destruction (Deleuze 1991: 27). It is this imperative and limitless limit towards and against which Sade repeatedly drives: it demands that his object of destruction be eternal, indestructible, and capable – over and over again – of suffering and torture. For Foucault, Justine must manifest an innocence which exhausts 'even the desire to torment it'; in the process she becomes the 'unattainable object of which she is the prime origin' (Foucault 1967: 284, 1970: 210). Serving a principle of perpetual immolation, as Dolmancé notes, is more than just a sexual or primary natural imperative, it is fundamentally political: 'one more effort', he urges the republicans, 'since you labour to destroy all the old foundations, do not permit one of them to survive, for let but one of them endure, 'tis enough, the rest will be restored' (Sade 1968: 301). One more effort and another and another: the sexual and political libertarian effort stretches onwards as absence of foundation and spectre of surplus.

Flash

'One more effort': Dolmancé's revolutionary injunction, like the acknowledgement of the relation between the shocks of revolution and the shocks of fiction, situates Sade and the tales of terror at a particular historical conjuncture. Their effects, then, are situated at the limits of discursive formations and shifts, at the limits of modes of writing, doubled and enfolded as transparency and reflection within and beyond the generic forms of repetition that reduplicate them; they also reverberate at and on the limits of the tortured body that neither cedes nor relinquishes its curious (non)objectivity. Modern literature starts in darkness and terror, and begins with a multiplication of mirror effects. It appears over an unformed abyss, a transitional zone where language is drawn 'by the overwhelming, the unspeakable, by thrills, stupefaction, ecstasy, dumbness, pure violence, wordless gestures' which open to a space of indetermination in which doubles, death, mirrors, and shadows of finitude proliferate (Foucault 1977b: 60). More precisely, as Foucault notes in 'Language to Infinity', it appears with 'Sade and the tales of terror', not simply as presentations of cruelty and evil, but disclosing 'something more obscure', something at the limit of language (1977b: 60). Literature, emerging amid the revolutionary shocks and darkness prefiguring Romanticism's vigorous imaginative, visionary, and fundamentally synthetic self-assertion of modern literary creativity, takes shape in modes of writing associated with violence and desire, in writing that is doubled between a language aimed at complete transparency, at the delivery of, as it were, unmediated effects, and a form that 'turns back on itself' that gives birth 'to itself in a play of mirrors that has no limits' (Foucault 1977b: 54). It is 'something like a mirror' (1977b: 54). But its reflexivity, severed from nature, reality, enunciation, context, and origin, refers to and reflects only itself as unbound literary utterance; it emanates in a virtual space of 'self-representation and reduplication', in the 'density of the mirror' (1977b: 56): this is not, Foucault spells out, the space of a reflected image, not mimesis, or, even, expression, but the very space of the double, the space of the black invisible and superficial line separating image and reflection: it is an 'infinite space where doubles reverberate' (1977b: 58). But if this reflection manifests a trajectory of virtual infinite regression, it simultaneously, breaking all representational parameters, broaches the outside. In Foucault's reading of Sade and the tales of terror, the role of the double – as both division and multiplication – splits the trajectory of writing between a superficial density of an ungrounded reflexivity and towards an almost unmediated transparency of violent effects: 'two twin and complementary languages were born at once the same central source' (1977b: 63). Moreover, birth occurs at a precise literary and historical moment with distinct and resonant forms and figures – mirror, monster, death, double – marking a point where language and literature overlap, separate, expend themselves in an infinite recession in which writing is severed from everything but returns with intense effects. Sade and the tales of terror evince what Foucault describes as the excess and the deficiency of language, an attempt – beyond utterance, literary form, formula and figure – to deliver the pure and violent transparency of shocking affect (terror, joy, thrill, stupefaction) that sees fiction bury itself ever deeper in the superficial density of reflection, repetition, and reduplication even as it tries to make itself palpably and immediately real. Among the metaphors that return, the figure of the mirror is exemplary of literary heterotopias, expending itself as real–unreal counter-site of representation, contestation, and inversion that goes on – expanding and receding – without limit, resolution, transcendence, or the finitude which gives birth to it. If the movement aims 'toward radical exhaustion' it succeeds in a multiplication and reduplication of playful expenditure (1977b: 62). Indeed, it is precisely in the space – the very thin line between image and object, self and reflection – that the double resonates as another critical figure of literary expenditure: a flattening of images to infinity is punctuated by a 'secret verticality' that engenders it, the 'narrow, black line which no perception can divulge' (1977b: 58). It is a space where transgression and limit, prohibition and excess discover their possibility.

The historical juncture at which Sade and the tales of terror appear, moreover, is significant in the manner that it opens to a particular relation of prohibition and suspension of law and structure, in the manner that, in the encounter with the unfixing of nature, idea, and word, it discloses a new relation of transgression and limit. Writing on Sade, Blanchot very elegantly catches the particular temporal spacing of revolutionary hiatus when the interruption of one order of prohibition ends before another begins. It discloses the 'monotony of its terrifying murmur' and distinguishes the revolutionary regime as 'pure time during which suspended history marks an era, that time between-times during which between the former laws and the new laws, reigns the silence of the absence of laws' (1967: 51, 56). It is, Blanchot goes on, an 'interval', '*l'entre-dit*' or 'between-saying', 'where everything ceases and everything stops' (1967: 56). Between saying – yet never fully said: not prohibition nor transgression but the thin line and abyss where limits flash and disappear, this space, not yet new in terms of law, subject, order, or literature, constitutes the murmuring absence out of which things may emerge, a space not of simple absence, limit, opposition, inversion, transition, or discontinuity, but where a language to infinity may appear in a flash, darkly, and brilliant in a moment of transgression. But not yet. It is not yet literature (though opening to the conditions of romantic vision): its mirrors, doubles, monsters, and intense effects may pull writing away from fixities of enunciation, mimesis, or representation and towards – through – the dense surfaces aimed at ultimate transparency, but the horrors of the abyss they reveal have yet to be ennobled, transcended, or occluded. Sade and the tales of terror appear at a specific, and brief moment, but one that is telling all the same: a moment, Foucault notes, 'when a language appeared that appropriates and consumes all other languages in its lightning flash' (1977b: 66). The flash of transgression, the pure and suspended time of their appearance, discloses only an internal fissure, a furrow, and the glimpse of an outside that is neither crossed nor closed.

Yet this is not quite the end of the play of mirrors, murmurs, flashes, and fissures. In 'Preface to Transgression' there is another flash in an essay that tracks the implications of Sade's writing for sexuality, the sacred, and subjectivity. 'Transgression', writes Foucault, 'is an action which involves the limit, that narrow zone of a line where it displays the flash of its passage, but perhaps also its entire trajectory, even its origin; it is likely that transgression has its entire space in the line it crosses' (1977a: 33–4). The essay – on the philosophical implications of Georges Bataille's writings – begins with Sade as a figure initiating the denaturing of sexuality for the modern age, opening it and nature up as an 'empty zone' and a 'fissure' 'which marks the limit within us and designates us as limit' (1977a: 29–30). This fissure, Foucault suggests, is the 'only division possible in a world now emptied of objects, beings and spaces to desecrate'; it is perhaps only a 'profanation without object' (1977a: 30). The rise of sexuality in modernity is, Foucault contends, an effect of the sacred having lost all positive character and meaning, with profanation only reiterating that loss. Following Friedrich Nietzsche, Foucault invokes the death of God to describe the effects of Sade's 'language of sexuality': it lifts us 'into the night where God is absent', addresses all action towards this absence but can only conjure up profanation which identifies, dissipates, and exhausts itself in that loss, in repeated – and vain – acts of transgression (1977a: 31). For Foucault, the death of God refuses the restoration of 'a limited and positivistic world', commanding only an exposure to an 'experience of limits, made and unmade by that excess which transgresses it' (1977a: 32). Following Bataille's assault on the philosophical subject, Foucault goes on to note how philosophy encounters another language it can neither master nor inhabit, a language in which it and its subject are displaced, dispersed, and multiplied, dragged through a labyrinth of loss and extreme expenditure and towards a spiral in which it dissolves: since Sade 'the universe of language has absorbed our sexuality, denatured it, placed it in a void where it establishes its sovereignty and where it incessantly sets up as Law the limits it transgresses' (1977a: 50). Depicted as a spiral, transgression, too, wanes. In its disclosure of finitude, in the ecstatic encounter with limit and death, the subject never sustains a sovereign position and the sacred never regains the full intensity of unproductive expenditure: 'The flash loses itself in this space it marks with its sovereignty and becomes silent now that it has given a name to obscurity' (1977a: 35). Transgression – once unleashing the violence of the sacred – cedes to a ceaseless profanation that only reiterates its loss. From there, to what and where? An abjection celebrated as a mark of identity? Vain horror courted as the possibility of feeling? What, indeed, becomes of the 'incessant murmur' that announces the mute, insistent recurrence of a Sadean and sub-literary impulse in Western culture: *120 Days of Sodom* turning into *50 Shades of Sade*?

Nightmare

Pierre Klossowski regards Sade's world as a 'vision of society in the state of permanent immorality': it is a 'utopia of evil' but one, he continues, that is 'paradoxical' in that it 'corresponds to the virtual state of our modern society' (1992: 62). Blanchot's prognosis for a contemporary vision informed by Sade is even starker: 'The world of the future will not be a world of values. Neither good nor evil will be its poles, neither virtue not vice, but the relationship to which affirmation and negation, carried to the extreme, correspond, by becoming identified with it' (1967: 54). Does this imply a permanent entre-diction, a ceaseless, yet banal and flattened between saying that is neither able to affirm nor negate laws, nor able to broach the intensity of sacrificial expenditure, replaying only the continual vacancy and vacation that Roger Caillois (2001) associates with the collapse of the sacred in late modernity? In this situation what happens to sub-literary space, to its murmur, mirrors, fissures, and flashes and to the limits it discloses? Indeed, do limits simply evanesce in a space without virtuality?

The interview in which Foucault recounts his nightmare concerns not only the writing of history and the invention of literature. It also makes a salutary observation on the discreteness and priority of philosophy, on its objects, reflections, and subjectivity: 'It seems that philosophy no longer exists, not that it has disappeared but that it has spread through a large quantity of diverse activities.' No longer, as in the nineteenth century, is it concerned with the 'possibility of objects', it is now evident in 'every activity which reveals a new object of knowledge, whether that activity comes under mathematics, linguistics, ethnology, or history' (1998: 292). Such a dispersal and dissipation transforms institutions and disciplinary boundaries as much as it reconfigures the circuits in which power/knowledge is produced. Perhaps, literature, too, pressed by creative capital and intellectual labour in the late twentieth century comes to enjoy a similar dispersal, its forms, and energies retooled and repurposed for neoliberal, post-industrial economies. In this context, literature is dispersed, following the fate of philosophy in the nineteenth century, to appear anywhere but 'literature', a disappearance calibrated to the rise of an economic principle in which all production, all poiesis indeed, is absorbed as a centre and surplus of an entrepreneurial and creative production of desires and information as much as goods and commodities (Gilder 1981; Goux 1998). The enterprising individual must creatively produce him or herself to live up to the neoliberal imperatives enshrined in the idea of 'human capital' (Foucault 2008: 225–6). If the expansion of an economic dimension forms one aspect of the postmodern condition set out by Jean-Francois Lyotard, another is the rise of technologies in the rewriting of what counts as reality according to optimal performativity rather than truth, reason, or morality: art becomes money as information and techno-economy reimagines the world (1984: 51–2). This is very much a post-literary trajectory: the vision, creativity, and inventiveness of literature dissipate and disperse into a wider neoliberal economy where they are put to work, serving as much as the imperative frames of circulation, as much as they are objects of production and consumption.

Curiously, the form of literature that Foucault identifies as subtending modernity resembles the dominant contemporary global medium. His account of the outside notes how literature develops, 'forming a network in which each point is distinct from even its closest neighbours, and has a position in relation to every other point in a space that holds and separates them all' (1986: 11). Networks, their points, propinquities, and distances, conjoined and distinct, compose a space, unsurprisingly perhaps, like the rhizomatic arrangement that displaces arboreal hierarchies (Deleuze and Guattari 1988). With a similarly strong and contemporary sense of the transformations of space, Foucault's discussion of heterotopia begins by hinting at some of the shifts being enacted by systems theory, informatics, and cybernetics, noting how space cedes to site, connection, and networked simultaneity: marking another 'epoch of space', the present of and in Foucault's lecture is envisaged as juxtaposition of distances and a world experienced as 'a network that connects points and intersects with its own skein', its dominant organizational figures being 'series, trees or grids' (1986: 22–3). This space of data, networks, points, and grid is not pursued, perhaps because it is not considered to be a heterotopia, perhaps because it had not, at that time, engendered a 'desanctification of space' and had not yet undone key oppositions (public/private; leisure/work). Foucault's lecture proceeds on the assumption that something of the sacred remains and people do not live 'in homogeneous and empty space' (1986: 23). Though concerned with the interrelation of virtual and real spaces when it comes to discussions of heterotopic effects, there is no development of questions of how networked virtuality might transform both the experience of space and the abyssal effects of literary and other heterotopias. What will virtuality do to that 'virtual space', in 'Preface to Transgression', where speech discovers its own resourcefulness, or when that 'virtual space of self-representation and reduplication' in which the 'density of the mirror' discloses the 'double of already doubled writing' (1977b: 55–6)? The enfolding of writing, of course, has different implications in the constitution and displacement of subjectivity to those of the entanglement of multiple links of a digital network, transforming – if not erasing – the narrow space of interrelation in which the transparency and density of language exert their effects. Where mirrors, flashes, fissures, and murmurs suggest density, division, darkness, and differentiation, respectively, current metaphors only emphasize transparency, like-ness, and transit: gateways, portals and, of course, ubiquitous windows realize the figures of fantasy fiction in an almost inevitable manner. It is a realization forgoing any late modernist irony of the kind to be found in Magritte's paintings of shattered glass: 'this is not reality', indeed, but the overcoding and supersession of any real. Transparency – of passage, communication and affect – is aligned with a new kind of density that eschews the tacit indeterminacy and heterotopic potential of the mirror, elides fantasy and reality into a new order of the fantastic that is neither utopia nor heterotopia, neither real or unreal but, in cruel mimicry of rhizomatic potential, marking a site of overwriting, one moving closer to a common sense of hyper-reality, and actualizing a conflation and obliteration of difference which only affirms pure, limitless, and frictionless virtuality, the endless potentiality of human capital flickering as the becoming of screens, the absorption of all perceptual spaces, coordinates and horizons in obedient like-ness, the muffling of all murmurs under the humming of the drive. Stop, stutter of tightened throat: an unreadable text, scrambled, indecipherable. It speaks; I like. Sleep on.

References

Blanchot, Maurice (1967), 'The Main Impropriety', trans. June Guinarchaud, *Yale French Studies*, 39: 50–63.

Caillois, Roger (2001), *Man and the Sacred*, trans. Meyer Barash, Urbana and Chicago: University of Illinois Press.

Canguilhem, Georges (2008), 'Monstrosity and the Monstrous', in *Knowledge of Life*, trans. Stephanos Geroulanos and Daniela Ginsburg, 134–48, New York: Fordham University Press.

Deleuze, Gilles (1991), *Coldness and Cruelty*, New York: Zone.

Deleuze, Gilles and Guattari, Felix (1988), *A Thousand Plateaus*, trans. Brian Massumi, London: Athlone Press.

Foucault, Michel (1967), *Madness and Civilization*, trans. Richard Howard, London: Tavistock Publications.

Foucault, Michel (1970), *The Order of Things*, London: Tavistock.

Foucault, Michel (1977a), 'A Preface to Transgression', in *Language, Counter-Memory, Practice*, trans. Donald F. Bouchard and Sherry Simon, 29–52, Ithaca and New York: Cornell University Press.

Foucault, Michel (1977b), 'Language to Infinity', in *Language, Counter-Memory, Practice*, trans. Donald F. Bouchard and Sherry Simon, 53–67, Ithaca and New York: Cornell University Press.

Foucault, Michel (1986), 'Of Other Spaces', *Diacritics*, 16(1): 22–7.

Foucault, Michel (1987), 'Maurice Blanchot – The Thought from Outside', in *Foucault Blanchot*, trans. Brian Massumi, 63–109, New York: Zone Books.

Foucault, Michel (1998), 'On the Ways of Writing History', in *Aesthetics, Methodology, and Epistemology*, trans. Robert Hurley, ed. James D. Faubion, 279–96, New York: New Press.

Foucault, Michel (2008), *The Birth of Biopolitics*, trans. Graham Burchell, Basingstoke: Palgrave Macmillan.

Gilder, George (1981), *Wealth and Poverty*, New York: Bantam Books.

Goux, Jean-Joseph (1998), 'General Economics and Postmodern Capitalism', in Fred Botting and Scott Wilson (eds), *Bataille: A Critical Reader*, 196–213, Oxford: Blackwell.

Klossowski, Pierre (1992), *Sade My Neighbour*, trans. Alphonso Lingis, London: Quartet Books.

Lyotard, Jean-Francois (1984), *The Postmodern Condition*, trans. Geoff Bennington and Brian Massumi, Manchester: Manchester University Press.

Radcliffe, Ann (1966), *The Mysteries of Udolpho*, ed. Bonamy Dobrée, Oxford: Oxford University Press.

Sade, D. A. F. (1965), 'Philosophy in the Bedroom', in *Justine*, trans. Austryn Wainhouse, 177–370, New York: Grove Press.

Sade, D. A. F. (1966), 'Reflections on the Novel', in *The One Hundred and Twenty Days of Sodom*, trans. Austryn Wainhouse, 91–116, New York: Grove Press.

Sade, D. A. F. (1968), *Juliette*, trans. Austryn Wainhouse, New York: Grove Press.

2

The living space of the image

Julian Reid

In its production space is subject to a critical division from which it emerges always as one of two kinds; either that of an inside or an outside. Space either interiorizes or exteriorizes, making insides and outsides out of what would otherwise be formless and functionless. What are the functions of interiorizing and exteriorizing forms of space? The fundamental function of interior spaces is that of the provision of shelter. The interior is the form of space which life requires for it to grow in a relation of security from forces located on and in the exterior. We encounter this basic assumption concerning the function of interiority directly in Henri Lefebvre's classic *The Production of Space*. 'The living being', Lefebvre writes,

> constitutes itself from the outset as an internal space. Very early on, in phylogenesis as in the genesis of the individual organism, an indentation forms in the cellular mass. A cavity gradually takes shape, simple at first, then more complex, which is filled by fluids. These fluids too are relatively simple to begin with, but diversify little by little. The cells adjacent to the cavity form a screen or membrane which serve as a boundary whose degree of permeability may vary. From now on external space will stand opposed

> to an internal space or milieu: here is the primary and most decisive differentiation in the history of biological being. The internal milieu will play an ever greater role; and the space thus produced will take on the most varied forms, structures and functions. (1991: 175–6)

This association of interiority with life, growth, and security permeates not just classical theories of space such as Lefebvre's, or the biological and evolutionary sciences from which they have often taken inspiration, but Western culture as a whole. The German philosopher, Peter Sloterdijk's three-volume *Spheres* project, describing itself as 'an archaeology of the intimate', is dedicated to excavating the origins, trajectories, and fates of the forms, both real and imaginary, of interior space which have, over historical time, given human beings the sense of their security in the world – a security which has permeated political and religious belief every bit as much as it has biological forms of thinking and practice. The modern age, however, as Sloterdijk describes, has been defined by a growing sense that the world in which we live, and on which we depend for our life, is not an interior at all. In the modern imaginary, the world is not an interiorizing space, which protects and encloses us, but an outside into which we are thrown, from the protective home we thought we had, amid the revelation of a false interior. 'Modern people have had to learn how one goes about existing as a core without a shell' and 'living in the modern age means paying the price for shellessness' as Sloterdijk expresses it (2011: 23–4). This task of learning has not inferred doing without interior spaces entirely. Instead it has demanded a different way of working the relations between inside and outside spaces. Where are we, Sloterdijk asks, if we are in the world? 'We are in an outside that carries inner worlds', he replies (2011: 27).

The shift which has taken place here, and which Sloterdijk directs us towards and describes, then, from the inside to the outside, is one that undermines our most fundamental assumptions concerning security. We no longer see the world as a space that avails itself to our protection. It demands that we form a different image of it; that is, an image of an outside as opposed to that of an inside. And indeed, the modern age has been defined, not surprisingly therefore, given its privileging of outside space, by a proliferation of images that draw and depict the un-homeliness of the world human beings inhabit. At their extreme these un-homely images become apocalyptic images that express the cunning of the modern era – 'the way it works against itself towards its own demise – the dreaming of its grave' (Calder Williams 2011: 16). These are images that have the effect of exteriorizing as well as being images of the outside as such. In this sense they are images of space, and of the one kind of space we moderns have privileged, over and at the expense of the interior spaces our elders once celebrated and cherished, as well as imagined dwelling within. They are images of what Gaston Bachelard, another great theorist of intimate and interior space, described as 'hostile space' – the form of space, that 'of hatred and combat' which he chose to ignore in describing their polar opposites – 'felicitous spaces' that 'may be grasped and defended against forces of adversity' (1992: xxxv–xxxvi). On the one hand, there is the house, an image which Bachelard privileged above and beyond all others as that singular image which gives human beings the sense of their security from the outside (1992: 17). On other hand, there is the universe and all the forces with which it besieges the house, and all the images through which we attempt to imagine our exposure to it.

The critical division of space, then, into insides and outsides, is made possible by an equally critical division of images. It is the imagination which in its seizing of space produces this division (Bachelard 1992: xxxvi). The division of space, the delineation of difference between interior and exterior spaces, always takes effect by way of the deployment of interior and exterior images. How is it that a space is able to interiorize, making of itself a form that belongs to the inside? It is precisely by way of the images with which that space becomes suffused by an imagination that is attracted to it (Bachelard 1992: xxxvi). Likewise the space which exteriorizes, and in doing so by which it constitutes the outside, can do so only by way of a deployment of images which arise from the repulsion of an imagination. Life, in its dependency on the creation of interior spaces for its security and growth, is as dependent therefore on images of a qualitative kind, as it is on those spaces. Does this mean that imagination is prior to and more fundamental than space in the cause of life? Absolutely. For while space requires the action of imagination, in production of the division between interiors and exteriors, in order for it to become useful to life, images of interiority and exteriority arise in independence from any prior properties of space.

*

In the text this book takes its inspiration from, *Des Espaces Autres*, Michel Foucault paid deference to the 'enormous work' of Bachelard for having given expression to the imaginal qualities of space (Foucault 1998: 177). As Foucault described, the spaces in which we live, including that often most ordinary and taken for granted of spaces, the domestic house, are not homogenous or empty but 'laden with qualities' and 'haunted by fantasy', Foucault recognized, in open debt to Bachelard (Foucault 1998: 177). Having expressed his deference to Bachelard, however, Foucault attempted to move, dramatically, in the opposite direction, by shifting his focus from the concern for internal and interior spaces that defined Bachelard's work to 'the space outside' (Foucault 1998: 177). Foucault asserted that this outside space is equally one *in which we live*. In being so, however, it is not a space which envelops us, protecting us, in order for our life to grow, but which draws us outside ourselves, erodes our life rather than protecting it, destroying our sense of time, by eating and scraping away at us (Foucault 1998: 177–8). These outside spaces are as equally suffused with imaginary qualities as their interior counterparts and the images that suffuse them possess qualities of as many differences and contrasts as those images which suffuse the interior spaces that Bachelard gave expression to. Foucault noted the importance of brothels and colonies as two examples of these kinds of outside spaces (or heterotopias as he also addressed them). In Chapter 1, Fred Botting details also the importance of literature as another space of the outside to which we moderns have gone, to disappear into its void (Botting 2018), and Anthony Faramelli, in Chapter 6, interrogates the importance of mirrors as such heterotopias (Faramelli 2018).

The tone in which Foucault addressed such spaces of the outside and their places in the origins and trajectories of Western modernity was of course very different to the melancholy of Sloterdijk and contrasted sharply and deliberately with the relative absence of their place in the analytics of space conducted by Bachelard. For Bachelard, to recall, the outside is simply 'the space of hatred and combat' (1992: xxxvi). It is where we go to die or at the very least risk death. For Sloterdijk too, the outside is a hostile space that necessitates the creation of interiors to provide us with a modicum of security. For Foucault, in contrast, the inside was never really the vaunted space of life it has been made out to be by these and many other authors. The biological, and therefore supposedly scientific, determination of the decisive importance of internal space, over and against the external, is precisely what Foucault's text functions to contest. As such *Des Espaces Autres* remains productive of a division within the theory not just of space, or of image, but of *life*. In positing the claim that the outside is where we actually live, in contrast with the interior spaces eulogized by Bachelard, still celebrated today by his disciple Sloterdijk, and asserted by Lefebvre, Foucault was effectively rejecting the assumption that life itself is fundamentally best lived and nurtured from the inside. The human, especially, Foucault ventured is absolutely not a being the life of which depends on the building of spheres for their own containment from the outside (Sloterdijk 2011: 28) but on a movement from inside to outside, and on an attraction to the outside which is every bit as decisive as the attraction which Bachelard credited the interior with.

We find this image of the outside as a space for the cultivation of life explicated forcefully in Gilles Deleuze's reading of Foucault. The human lives not simply by retreating from the outside, or protecting itself from the outside through the constitution of an inside, but by entering into relations with new forces from the outside (1988: 126–7). As Deleuze asserted, this way of conceiving the space of life is not antithetical to biological science. For this is what the biological process of mutation entails, an entering into relation of forces within with forces outside the living being. An encounter which takes place outside of what life commonly calls its 'comfort zone', but which, in spite of our attachments to comfort zones, so well described and deconstructed by Culley in her contribution to this book (2018), we understand as being necessary for our growth. An uncomfortable confrontation and struggle must occur on the outside and with the outside in an act and process of bringing within. The living being must fold into itself that which it encounters outside of itself. This is how form as such emerges, through the folding of forces of the outside into the inside. Form always results from a new relation between these forces (1988: 132) and the human becomes more powerful to the extent that it is capable of taking charge of the very processes by which its form is thus constituted. Its power does not result from a retreat or defence from the outside but through the active encounter with and selection of forces from the outside, and the making of itself, through a folding into itself, of those forces it so selects.

* *

It is remarkable, then, that there exists no equivalent study of the poetics of exterior spaces to that of Bachelard's analysis of the poetics of interior space; no phenomenology of the images through which we construct exterior space as felicitous and attractive. Not least perhaps because the idioms through which the outside is imagined remain heavily geared towards its construction as hostile and threatening. A full exploration of the poetics of exterior space would have to tell a different story and produce a different image. That said, *Des Espaces Autres* is not without its images. The text ends indeed on a very particular image, neither that of the brothel nor of the colony, but the ship; that 'floating space' and 'placeless place, that lives by its own devices, that is self-enclosed and, at the same time, delivered over to the boundless expanse of the ocean' and which while enclosing and interiorizing us within its bounds delivers us over to the boundlessness of the sea, so that we might travel from port to port, brothel to brothel, colony to colony, and so on (1998: 184–5).

The ship is the space of imagination par excellence in this sense, Foucault maintained, being one in which we live, and which encloses us, protecting us from an outside which will otherwise swallow us, while delivering us to outside spaces which together constitute the voyage (Foucault 1998: 185). It is the space in which these two regimes of the image coincide, and in which imagination experiences the subtle confluences of images of the outside and the inside as they coincide, coexist, and constitute a space in which both they and we live together. It is in other words, neither simply an image of a space nor an image of imagination as such, but space-image.

What is space-image? It is not an image of space, nor simply a spatialized form of image, but a space in which images live, that is, the living space of the image. It is a space in which the imagination locates itself as well as an image in which the imagination sees itself, that is, the image of imagination. For imagination can only live in particularized kinds of images. It is not every image that can be a habitat for the imagination. In fact imagination abhors the vast majority of images, finding no space there. *Des Espaces Autres* was not the first time that Foucault had discussed and asserted the importance of the ship as both space and image. *History of Madness* opens with a similar description of the poetics of the ship of fools; that 'strange drunken boat' on which the mad of the Renaissance would be interned and on which they would drift from one port to another (Foucault 2006: 8–12). These ships were, in equal proportions, real and imaginary spaces. They existed in the reality of the Renaissance while also occupying a 'privileged place' in its imaginary landscapes being a 'literary commonplace' of the time (Foucault 2006: 8–9). Their symbolic importance arose from this precise capacity to transcend the binaries of inside and outside spaces, and to be the space in which these two different regimes of the image could coincide, and thus through which the imagination could locate itself, in an image – an image that creates a space in which imagination *can actually live*. It is its being the image of imagination as such which gives the ship of fools its special symbolic value. The madman aboard the ship of fools is placed on the inside of outside space as much as he is on the outside of an interior (Foucault 2006: 11).

This is not, then, a space, simply, *in which we live*. It is a space in which imagination lives; the living space of the image, and not of human beings, with their extra requirements. This idea, of such a space, in which imagination and the images it creates may live, is not revealed or expressed explicitly in *Des Espaces Autres*. Foucault's explication of the relation between life and space in that text revolves simply around the question of the spaces *in which we live*, and the text constructs a simplistic binary relation between outside and interior spaces in the context of which he chooses to privilege outside space as the space *in which we live*. The ship is, in fact, however, neither an outside nor an inside space. It is another kind of space, one that does not concern our life as human subjects entirely but the images of which we are imperfect homes. We are, as the artist Bill Viola has asserted and emphasized in his work, 'living databases of images – collectors of images' which never stop growing and transforming within us once inside us (quoted in Agamben 2013: 5). The human is that peculiar house which images inhabit, and through which images are given time, and thus life. It is the space which interiorizes, and on account of which images are protected, from their outside, the diminishing space of reason which would otherwise gladly subject them to its excoriating gaze.

The human, then, is a home to images, a space in itself containing and protecting the image from forces that would otherwise destroy it. It is the vessel in which images voyage upon their outside space, the sea of reason, which would otherwise swallow them; a felicitous space, the human, without which images would drown. But the human who becomes the vessel to this cargo is himself at risk. It is a fact of course that the human beings who inhabited the ships of fools in which imagination found its sanctuary were mad subjects, whose madness was said to arise from the affectations of their imaginations. As Foucault described imagination was typically identified throughout the history of madness as the faculty most at risk of being affected by the malady, and thus, that which is most at risk in its onset. Hallucinations, for Boissier de Sauvage, writing in the eighteenth century, were 'sicknesses whose principal symptom is a depraved and erroneous imagination … errors of the soul occasioned by the vice of organs situated outside the brain, which allow the imagination to be seduced' (quoted in Foucault 2006: 197). Such 'troubled and strayed' journeys of imagination, when coupled with corporeal disturbances, were said likewise by practically every early modern theorist of madness to be accountable for 'delirium' (Foucault 2006: 198). But as Foucault stressed in his study of early modern diagnoses of madness, within these theories, the imagination itself was considered to be innocent (2006: 232). Images, in themselves, were not thought to be mad, nor were they in themselves thought to cause madness. 'Madness begins' only when a subject 'lends a value of truth to the image' concerned (2006: 232). Just as the sane subject exercises its sanity by refusing to provide them with the hospitality they seek, by confronting images, limiting them, and interrogating them from a distance, so the mad subject is made mad not by the image in itself, but by 'turning all that is given in the image into an abusive truth' (2006: 232). Madness, therefore, was said to entail, a going beyond images, by 'allowing the image to take on the value of total and absolute truth' with the mad subject nevertheless remaining embedded *inside* the image (2006: 232). The sane subject maintains its sanity by stepping outside of the image and perceiving its unreality while the mad subject, by contrast, 'allows himself to be totally caught up in its immediate vivacity', becoming 'entirely absorbed in it … inside the image, trapped within it and incapable of escaping' (2006: 232). Sanity, in other words, is not preserved by divesting oneself of one's images, but by the act of stepping outside of the space of the image, subjecting it to reason, with a view to seeing its limits, while madness entails an act of stepping inside, and looking at the world from within the space of the image, affirming its truth from within, by a becoming at one with the image, in the act of hospitality which images demand. Either one becomes the space in which images live, space-image, or one refuses them refuge. But in the context of this decision as to whether to step out or into the image, images, and the faculty for their production, which is to say, imagination, remains innocent. The world of imagination, then, it is crucial to recognize, has been conceptualized in practically all modern theories of madness, principally, as a space – a space into which one either steps in or steps out, but the relation with which one cannot avoid. Reason was likewise conceptualized as that technology which enables one to assume the necessary distance from the space of images.

* * *

In describing the space in which images live we describe a space in which modernity has attempted to define the limits of, to affirm the innocence of, while asserting the dangers it poses once the human allows itself to become. Giorgio Agamben, in his classic *Homo Sacer*, used the Mobius strip to describe precisely this kind of space, in which outside and inside pass through one another (1998: 37). Unlike the privileged space of the ship, however, it is a space which we struggle to see, to make an image of, because it remains hidden from our eyes, and which for that precise reason, Agamben calls upon us 'to fix under our gaze' (1998: 37); To behold this space, where inside and outside pass into each other, and to make an image of this space, which is neither inside nor outside, but the two coinciding; to see it, neither as one nor the other. In his seminar on *Anxiety* Jacques Lacan described the movement of an ant upon the surface(s) of a Mobius strip, and how in passing over from one apparent face to the other, the ant does so, without ever going 'over the edge' (Lacan 2014: 96). Is this challenge of seeing this space that Agamben poses for us, not the challenge of becoming the ant which, walking from one surface of this space to the other, remains oblivious to the fact that it is passing from one to the other? The challenge of occupying, territorializing, and moving upon and within this space in ways that will entail us in never going over the edge? This challenge is not only both these things but also one of forming precisely an image which is without interior or exterior qualities. It is neither the image of the inside nor the image of the outside. It suffuses neither an interior nor an exterior space. Nor is it an image with the qualities we associate with interiorities or exteriorities. It is not an image that protects every bit as much as it is not an image that subtracts or threatens. It is the image we must suppose which the ant possesses, of that surface of the strip, which to us we can only behold in the manner of a space which is difficult to conceive, and which we fear, were we to attempt to traverse it, will send us over the edge. A space which we cannot occupy or hold in common without recognizing the realities of its edges.

This space we cannot hold, this inhospitable space, is the very space in which images live, and without which they cannot live; a space where, living within, the image also lives, entirely on the outside (Blanchot 1993: 321). As such the human who is predisposed towards images, their worlds, and their lives cannot but be made ill by the will to be their host. One can choose, as each of the theorists of the space of the image we have encountered so far, did, between interior and exterior regimes of imagination and remain within the relative safeties of the districts which interiority and exteriority avail. Or one can do as perhaps only what Blanchot urged us to do, and 'approach through the image, the very space of the image, the outside that is its intimacy' (1993: 324). It is not a question of choosing between interiors and exteriors. Instead it is a question of negotiating the relation between the two spaces of image I have only schematically sketched the nature of here: the interior and the exterior. Moreover, it is a question of understanding how these two spaces in which images live function always within the same image at any one time. In other words, it is a question of addressing the inherent dangerousness of images with a view to comprehending how the life of the image functions in a consistent movement between interior and exterior spaces. Without images, without an understanding of the movement that occurs between these two spaces of imagination especially, we cannot hope to comprehend the ways by which images move us between these two forms of space. Images are not, as Lefebvre would have us believe, simply 'fragments of space' empowered with the abilities to fragment space itself (1991: 97). It is not 'the image' that 'kills' (Lefebvre 1991: 97), for there is no homogenous thing such as 'the image'. Our reception of and movement within spaces themselves are dependent upon images, the ways in which they meet our gaze. The eye is not averse to tactile experience. The problem is to parse the differences between images that kill space and images that create space.

* * * *

Bracha Ettinger approaches an account of this problematic in her development of the concept of *fascinance* (2006). *Fascinance* is what happens to the gaze when it encounters an image that gives life rather than taking it. It is in that sense the opposite effect of what Lacan described in terms of the *fascinum*. It is the transformational aesthetic effect that occurs in the subject when his or her gaze is returned by the image that fascinates (Ettinger 2006: 61). In the fascinum that Lacan described, the gaze is not returned, for whatever reason, and the subject is determined, stopped in its movement, by this event of non-reciprocity of the image. In the case of fascinance the image puts the subject into movement, and optimally, transforms it. To explicate her case Ettinger gives a reading of *The Ravishing of Lol Stein*, a novel by Marguerite Duras (1966). The transformation which the image in the novel possesses the power to perform is one from girlhood to womanhood – that of Lol, the novel's protagonist. But Lol's movement is arrested and the rest of the novel tells the story of her attempts to recreate the same or a similar encounter with an image by which she might complete her passage, by turning *fascinum* into *fascinance*, so to speak (Ettinger 2006: 62–3). An encounter with an image whose transformative potential has failed, and the attempts to recreate a comparably endowed image, in order to escape from fascinum into fascinance, is, Ettinger argues, the story of the ravishing of Lol Stein.

How can an image affect such a transformation? How can an image deny its own potential to transform? In the case of Lol, her potential transformation from girl to woman is located in an image of two lovers, a man and a woman. It is for Lol, 'a fascinating image she wants to see again and again' (Ettinger 2006: 73). An image not just of a man, but a man who 'in front of the desiring and fascinated gaze of Lol', the girl, is desiring another female figure, a woman, a mother figure, who likewise is fascinated and desiring the man, in representation of the image of Lol as the woman she actually desires to become (Ettinger 2006: 73–4). Lol is, Ettinger argues, not yet ready, as a girl, to love the man in question, and desires only to gaze at the image and become a part of it, 'by the transgressive force of fascination that will include her' (Ettinger 2006: 64). But how could such an image possibly perform such a transformation? What is the nature of this 'transgressive force' that Duras portrays and Ettinger identifies in the image in question? What quality differentiates it from the image that performs the opposite effect? For Ettinger, it is a question of difference not of the image as such, but of affect. This power and difference in affect infiltrates the field of perception and shapes the image such that it becomes, either the purveyor of *fascinance* or *fascinum* (Ettinger 2006: 67). Either the image welcomes her into it, making its space 'shareable', Ettinger maintains, or it will shut her out, closing its space, fragmenting the potential by which a space in which she, Lol, is implicated, might have become shared. But the image, here, is merely a shutter, which either opens or closes, upon a space, depending on the operation of an affect. The image itself is dead matter, neither animate nor inanimate, immobile, until it is moved, by the power of affect. The gaze of *fascinance* is, if we follow Ettinger, 'an affective vibration' (2006: 85). It allows a 'glimpse at the forever out-of-time-and-space' by which 'the forever-future and the archaic past join in the now of the co-eventing in the Real' and 'transform the old scar or mark' (Ettinger 2006: 85).

What is the 'forever out-of-time-and-space' if not the time-image which we encounter in the philosophy of Deleuze (1989; 2005)? The image, that is to say, of time itself? Is it not also the small non-time-space in the very heart of time that we also encountered in the work of Arendt (1993: 13)? But in Deleuze's case we remember that the encounter with this image does not enable movement, instead it disables. But the disablement is only superficial. For the encounter with time-images deepens us, taking us down, from the surfaces of movement on which we are otherwise positioned, in our sensory-motored lives. In doing so it makes us a subject. For subjectivity itself emerges, not in movement as such, but in the gap between a received and an executed movement (Deleuze 1989: 47). It is not, as Deleuze argues, 'motor or material, but temporal and spiritual' (1989: 47). But what is crucial in Deleuze's treatment is that it is the image that generates affect. Affect does not simply enter, infiltrating a field of perception, shaping an image; it is instead the encounter with an image, which produces the affect, and the image of time especially. Affect, according to Deleuze, belongs to the gap, between stimulation and response. In this sense affect inhabits an image, because it is an image that fills that gap. What Deleuze called the recollection image (1989: 47).

In terms of its movement within time the exterior image is also more complicated than Ettinger's analysis allows us to understand. It is not simply initiatory of a movement from girl to woman in the manner she argues it to be, for it is the cause of a forward backwardness which deepens our relation with the past every bit as much as it renders the future more shallow. The Japanese novelist, Yukio Mishima, understood and conveyed this in his *Runaway Horses* (2000). There we encounter the character Honda, whose mnemic-images of his childhood friend, Kiyoaki, produce a movement that is not from boy to man, but man-boy. When he was young there was but one reality for Honda, while the image of the future was multiple, 'swelling with immense possibilities' (Mishima 2000: 7). Now he is old reality takes many forms, and it is the past which seems 'refracted into innumerable possibilities' (Mishima 2000: 7). Each of his images of his childhood friend, Kiyoaki, contains its own reality, and effectively annuls the real with its weight of presence. Inside of it he moves backwards, in time, opening the past, fragmenting it into its innumerable possibilities, transforming from man to boy. Images of what had occurred became constitutive of what could have occurred, and what could have occurred takes on the form of the more vivid than real. Is this the recollection image that Bergson spoke of and which Deleuze also theorized (Deleuze 1989: 47)? No, this is not simply an image that recalls some scene that happened in the past; for this is an image which operates as a function of the future by giving the past all the multiplicity and possibility which futuricity entails. It is not simply that the future becomes impoverished, narrow, on account of the potency of the image of Kiyoaki for Honda. Instead the future gives what it has, in terms of its multiplicity and possibility to the past, such that the past becomes future-full. Honda's future retains the images of Kiyoaki, apparently drawn from the past, such that they make them into 'objects to come' (Deleuze 1989: 52).

This is the paradox of the exterior image, and its relationship to movement. It enables an ascent to subjectivity, but it is an ascent that proceeds downwards. This is not to be confused with the movement of descent that Bachelard described when he addresses 'the imaginary fall' (Bachelard 2002: 91–109). Indeed the somewhat simplistic differentiation between ascending and descending movement upon the vertical axis of imagination that Bachelard proposes has to be revised accordingly. It is not simply that there is an exterior axis of imagination that is true and an internal axis of imagination that is false (Bachelard 2002: 92). Bachelard proposes that upwardness is the only true direction that the imagination is geared towards, while downwardness is a direction we take when present in the order of the real. In contrast, one falls high every bit as much as one ascends down. The upward fall is what characterizes the encounter with the aerial image every bit as much as the downward climb characterizes the abyssal image. An upward climb would qualify only the most banal of images, as much as the downward fall.

Nor is this movement initiated by the aerial image merely temporal or spiritual in the manner that Deleuze argues it can be understood, once we do away with a motorized or material understanding of movement. It is also political. It is a movement that goes back as much as it goes forwards with a view to operating upon the present, and transforming the real by acting upon it, impregnating it with images drawn from the past, which now takes on the future, in a movement which dispels nostalgia, conquers melancholy, is neither happy nor sad, but affective. 'So it is that time reenacts the most curious yet earnest spectacles within the human heart. The past makes its appearance again, with all its mingled dreams and aspirations' (Mishima 2000: 29). This spectacle, of the past made present, future-full with all its dreams and aspirations, ready, now to be realized, is that of the image of a movement by which the subject climbs downward, falling upward, forward back, through time, and into new spaces. So it is that images which appeared dead come back to life and charge us with a life which brings past and future into a relay in the present.

* * * * *

There is no image of space nor space of images except the living space which subverts the very division on which space depends for its human comprehension, settlement, and exploration. The space co-substantial with the life of images is neither that of the interiority explored by Bachelard and still eulogized by his disciples including Sloterdijk nor that of the exterior championed by Foucault. It is the inside–outside which theorists of space and images come close to, now and then, without being able to either penetrate or escape from. It is an imageless space, or at least a space for which no human image suffices. It is a space which reveals the impossibility of reconciling relations between human life and imaginal life. The space of the image attracts but it is a space which human cannot hold. It is a space of illness, and yet, also, of felicitous affect, allowing a glimpse of the forever out-of-time-and-space which particular images, and what Deleuze described especially as time-images, reveal; the space, that is, where subjectivity is made, as much as human life is lost; the living space of the image.

References

Agamben, G. (1998), *Homo Sacer: Sovereign Power and Bare Life*, Stanford: Stanford University Press.

Agamben, G. (2013), *Nymphs* (London: Verso).

Arendt, H. (1993), *Between Past and Future*, London and New York: Penguin.

Bachelard, G. (1992), *The Poetics of Space*, Boston: Beacon Press.

Bachelard, G. (2002), *Air and Dreams: An Essay on the Imagination of Movement*, Dallas: Dallas Institute Publications.

Blanchot, M. (1993), *The Infinite Conversation,* Minnesota: Minnesota University Press.

Botting, F. (2018), 'Literature-Outside-Space: Foucault, Sade, and Tales of Terror' (this book), in R. G. White, A. Faramelli, and D. Hancock (eds), *Spaces of Crisis and Critique: Heterotopias Beyond Foucault*, London: Bloomsbury.

Culley, M. (2018), 'Inside Comfort: The Interior and the Immune System' (this book), in R. G. White, A. Faramelli, and D. Hancock (eds), *Spaces of Crisis and Critique: Heterotopias Beyond Foucault*, London: Bloomsbury.

Deleuze, G. (1988), *Foucault*, (London: Athlone Press.

Deleuze, G. (1989), *Cinema 2: The Time-Image*, London: Athlone Press.

Deleuze, G. (2005), *Cinema 1: The Movement-Image*, London and New York: Continuum.

Ettinger, B. (2006), 'Fascinance and the Girl-to-m/Other Matrixial Feminine Difference', in Griselda Pollock (ed.), *Psychoanalysis and the Image*, 60–93, Oxford: Blackwell.

Faramelli, A. (2018), 'Mirrors and Masks: The Political Space of Zapatismo', in R. G. White, A. Faramelli, and D. Hancock (eds), *Spaces of Crisis and Critique: Heterotopias Beyond Foucault*, London: Bloomsbury.
Foucault, M. (1998), 'Different Spaces', in Michel Foucault (ed.), *Aesthetics, The Essential Works 2*, 175–86, London and New York: Allen Lane.
Foucault, M. (2006), *History of Madness*, London and New York: Routledge.
Lacan, J. (2014), *Anxiety*, Oxford: Polity.
Lefebvre, H. (1991), *The Production of Space*, Oxford: Blackwell.
Mishima, Y. (2000), *Runaway Horses*, London: Vintage.
Sloterdijk, P. (2011), *Bubbles: Spheres 1*, New York: Semiotext.

3

Inside comfort: The interior and the immune system

Sheena Culley

Fashioning a shell

For Gaston Bachelard, the shell, alongside the nest, the corner, and the drawer, chest and wardrobe, represents safety and protection. 'Everything about a creature that comes out of a shell is dialectical' (1969: 108). Bachelard was inspired by Paul Valéry's writing on the shell, where he reads the process of building as more than protection. 'One must live to build one's house, and not build one's house to live in', he writes (106). On picking up the sea shell, Valéry not only contemplates its form, he imagines how one might go about creating such a dwelling place. Through these observations, Valéry begins to connect creation, being, and subjectivity. How should one go about fashioning a shell? (120). The mollusc who makes the shell involves itself in the process of creation with neither blueprint nor design. The creature has no idea in his mind, and requiring neither knowledge nor memory, proceeds to habitually fashion his house. Taking his raw materials of calcium salts, which are ingested and pass into the blood stream, the mollusc exudes its 'epithelium', its outer layer. Valéry writes:

> This gives us the outside of the shell. But it grows in thickness, and this growth involves very different material, structure, tools. Protected by the solid rampart that the edge of the mantle has built, the rest of this admirable organ fashions the refinements of the inner wall, the water-smooth lining of the animal's home. There is nothing too precious or delicate for the meditations of a life so much of which is spent at home. (1977: 129)

The act of building the shell has no meaning to 'man's little intellectual sphere' (Valéry 1997: 127), and this is what makes the shell such an object of curiosity for Valéry. Relating the activities of this organism to the human, Valéry comes to imagine the mollusc as a private being, dwelling in his 'den', 'fortress', or 'masterpiece'. He must confront two different 'realities' or 'geometries'– that of the outside and that of the inside of his dwelling place. 'Perhaps', he playfully suggests, 'he measures his private "time" by the sensation of secreting a little prism of calcite and putting it in place' (133).

Valéry's 1936 observations of the mollusc are as much a reflection on culture as they are on nature, evoking a relationship between the interior and exterior and the public and the private, which characterized nineteenth-century dwelling. In *The Arcades Project*, Walter Benjamin made a connection between the act of fashioning a shell and nineteenth-century bourgeois dwelling. He wrote, 'The nineteenth century, like no other century, was addicted to dwelling. It conceived of the residence as a receptacle for the person' (2002: 220). This receptacle was replicated through many objects of encasement that characterized material culture of the time:

> What didn't the nineteenth century invent some sort of casing for! Pocket watches, slippers, egg cups, thermometers, playing cards – and in lieu of cases, there were jackets, carpets, wrappers, and covers. (220)

Benjamin compares the bourgeois interior to 'the inside of a compass case, where the instrument with all its accessories lies embedded in deep, usually violet folds of velvet' (220). The trace of the subject is left as an imprint, moulded by repetitive, habitual habitation:

> 'To dwell' as a transitive verb – as in the notion of 'indwelt spaces'; herewith an indication of the frenetic topicality concealed in habitual behaviour. It has to do with fashioning a shell for ourselves. (221)[1]

The problem with nineteenth-century dwelling for Benjamin was the difficulty in seeing it as a historical condition, as the idea of dwelling conjured up images of the 'eternal'. Rather than Valéry's shell, for Benjamin, it was the maternal womb which featured as the most primal of these images. Its nineteenth-century material representation came in the form of upholstery and padding which served as a reaction to modern city life, comforting the inhabitant from the shocks of new technologies which produced 'alienating forms of experience' (Rice 2007: 10). An historical bourgeois identity was created under the mask of an eternal subject.

Dwelling in comfort

Benjamin's and Valéry's musings on nineteenth-century dwelling, survival of the organism and subjectivity emphatically bring together some of the central ideas we hold about comfort today. Comfort is seen as a form of protection, enclosing the subject from a hostile exterior. Comfort is an immune system, which protects the interiority of the subject from the hostility of the outside world. The trajectory of literature on the topic of comfort references domestic space and the home, defining comfort in the terms of the Victorian ideals that Benjamin commented on. John Gloag's *Victorian Comfort: A Social History of Design from 1830–1900* (1973) traces comfort through domestic spaces, objects, and practices. Gloag highlights the importance of bay windows, net curtains, ornaments, antique furniture, fireplaces, and bloated sprung armchairs that played a central role in defining comfort as a middle-class dream of 'exclusive privacy', protecting the subject from the dangers of the outside world. However, Crowley argues that comfort, a seemingly 'natural' idea, is instead socially and historically constructed (1999: 750). The main focus for Crowley is the shift from the feudal system to capitalist production, and the changing attitudes towards consumption that resulted from this shift. It was during this time that comfort became bound up with ideas of 'emulation, refinement, self-fashioning, conspicuous consumption, and romantic illusions' (753). Daniel Miller's *The Comfort of things* (2008) is an anthropological study of domestic space, based on a London Street. The first two portraits, 'Empty' and 'Full' support Dickensian visions of comfort associated with cosiness and conviviality, where material objects such as furniture and home cooking portray an idealized form of dwelling.

Whereas historical and contemporary texts on comfort focus on domestic space, this chapter explores a parallel history of comfort that is concerned with another history of space so to speak – a cultural history of the modern body. While many of comfort's historians acknowledge that the body plays a vital role in our understanding of comfort as a physical and material discourse (e.g. Crowley defines comfort as a 'self-conscious satisfaction between one's body and its immediate physical environment' (1999: 750)), histories of bodies are absent from their historical explorations of comfort. By approaching comfort from the body, this chapter explores an alternative delineation of comfort: comfort as an immune system, which brings together ideas relating to nineteenth-century dwelling and nineteenth-century bodies. While immunity does not etymologically refer to 'protection from', this definition is taken as a point of departure, as it is what we understand and imagine both immunity and comfort to comprise of in everyday terms. The power of metaphors and 'poetic borrowing' between science and culture is fundamental to our understanding of modern bodies, and important in how we have come to understand comfort.

The privileging of interiority that has come to define comfort has little to do with the original meaning of the term. The English word *comfort* comes from the Latin *confortāre*, meaning 'to give strength' or to strengthen (*-fort* meaning strong and *con* meaning with, but also acting as an intensive prefix) (*OED*: 534). *Con* later became *com* via phonetic change as *confortāre* passed through Old French (*confort*) to Middle English, and became *comfort*. The verb *comfort* has a wide range of meanings which are now obsolete. Most closely related to its Latin roots, comfort meant to encourage, or to inspirit. This sense of encouragement can be seen in terms of providing physical strength or support, as well as to 'invigorate' or 'refresh'. In addition, comfort was thought of in terms of relief, either by means of assisting in sickness or affliction, or to 'soothe in grief or trouble', or in providing 'relief or support in mental distress' (*OED*: 533). This meaning is indeed still used today. Comfort therefore carries a strong sense of well-being over anything overtly interior or even spatial by definition. Due to its association with well-being, comfort is inextricably linked to bodies. There is a belief that comfort in pre-modern times primary referred to a spiritual state, which was then transposed onto material culture in a bid to justify increasing standards of living (Crowley 1999: 750–1). However, as is apparent from the range of definitions from the thirteenth century, comfort has always been about 'physical' well-being. As Jacques Pezeu-Massabuau states, 'There could never be a theory or use of comfort that does not begin from the body, the group of cells from which we are made and upon which our well-being rests. Everything that motivates or limits comfort stems from the body' (2012: 21). By drawing on the biopolitical notion of immunity, and the idea of comfort as well-being more generally, it becomes apparent that the modern body, in particular the notion of boundedness, has played a particular role in defining comfort as a phenomenon of interiority and protection. Following the work of Ed Cohen, I suggest that specific moments in his genealogy of immunity, which imagine the body as a discrete, bounded object opposed to a hostile exterior, also played a role in defining our present-day understanding of comfort. These ideas, as Cohen demonstrates, are never merely a project of biological thinking; they are always biopolitical. This, I argue, opens up a connection between defining comfort in terms of domestic space and bodily space, as bodies come to be imaged as property. Another late-nineteenth-century science also drew heavily on the image of the bounded body: psychoanalysis. I argue that psychoanalysis, in particular Freud's biological imagery, has also contributed to defining comfort as an immune system, and as such, defines discomfort as trauma. I finally consider whether we can think of comfort outside of these Freudian terms.

Science, culture, and modern bodies

Elizabeth Grosz reminds us that there is no separation between a real, material body and its cultural and historical representations. Our conception of bodies is informed by the physical sciences and shaped by medical, chemical, and biological knowledge (1994: x–xi). As a result, 'the' body lives as an idea historically constituted by wider systems of belief that shape the epochs that we inhabit. The modern body has become a metonym for the subject, a bounded, individuated unit. Of the medieval, grotesque body, Mikhail Bakhtin writes, 'It is unfinished, outgrows itself, transgresses its own limits.' 'The limits between the body and the world are erased', leading to the fusion of the one with the other and with surrounding objects (1984: 26, 310). In contrast to the open, pre-modern body, the bodily functions ('sexual life, eating, drinking and defecation') of the modern body bear no relation to social life and instead inhabit a 'private and psychological level' (317). As a result, 'all orifices of the body are closed' (320). Rather than belonging to the continuous cycle of life and death, the modern body produces itself, and the human world 'creates itself from itself', rather than emanating from a divine and mysterious principle. Ferguson writes, 'Free from earlier forms of subjugation, modernity comes to itself in perpetual inner motion, as a continuous process of restless self-production' (2000: 3). Central to our understanding of the modern body is not only the idea of autonomy, but an array of dualisms, including subject/object, self/other, and Ego/world. The body, since Descartes, has been spatially located, seen as a subjective inside opposed to an objective outside. Therefore, hardness and impenetrability, as well as space-filling extension have become central to the way in which we visualize it (6–7). Ed Cohen states that 'the modern body aspires to localise human beings within an epidermal frontier that distinguishes the person from the world for the duration which we call life' (2009: 7), and 'the human organism imagines itself as distinct from an environment that only subsequently seems to surround or even oppose it' (10). It is this modern, bounded body that is common to defining both our understanding of corporeal comfort and immune systems, binding these two ideas together.

Immunity and modernity

By adopting a Heideggerian perspective which defines dwelling as building, Peter Sloterdijk theorizes inhabiting the modern world as constructing immune systems. Sloterdijk's work makes an important connection between immunity and comfort, although he does not refer to comfort as such. In the first volume of his magnum opus, *Spheres*, Sloterdijk describes the modern age as 'shelless', arguing that modernity since Copernicus has displaced humans from their cosmological locus as the sun displaced the earth as the centre of the universe, exposing us to a new outside. Mankind is 'towered above on all sides': 'Taking part in modernity means putting evolved immune systems at risk' (Sloterdijk 2011: 23). The shattered illusion of our home's central position in space has deprived us of 'the comforting notion that the earth is enclosed by spherical forms like warming heavenly mantles'. Science, the modern form of truth, came to replace religion, the death of which also plays a major role in exposing the human being. As a result, the moderns 'have had to learn how one goes about existing as a core without a shell' (23). The bursting of 'God's shimmering bubbles' leaves the modern being exposed and shelless; he preoccupies himself with creating spheres, protecting himself from the abyss of infinite space (24):

> Now networks and insurance policies are meant to replace the celestial domes; telecommunication has to re-enact the all-encompassing. The body of humanity seeks to create a new immune constitution in an electronic medial skin. (Sloterdijk 2011: 25)

Whereas Sloterdijk names this phenomenon as the 'new immune constitution' or 'the building of spheres', spheres could indeed be defined as the modern phenomenon of comfort. Spheres stop us questioning where we are, instead allowing us to focus on who we are. Not limited to technological or social practices, psychoanalysis forms part of the modern immune system, which Sloterdijk argues plays a particularly pertinent role in replacing the question of 'where' with 'who' (2011: 30). While Sloterdijk's theory of spheres is not one of enclosed immunity, or isolation, where no mediation takes place between the subject and the world (this would be defined, according to Sloterdijk, as existing as a 'monadic ego orb' (86)), boundaries remain important to his theory. Spheres are immune systems opposed both to creating isolated bubbles and to militaristic defence from a hostile exterior. It is this definition which therefore forms the basis to view comfort as an immune system, noting the importance of boundaries (albeit interacting in an intersubjective fashion). In fact, Sloterdijk's conception of immunity contains within it immunity's own antonym: community, a point we shall refer to later.

Rather than being defined by intersubjective dwelling, immunity, when thought of in biological terms, has historically been conceived of as a form of protection, one which cuts an organism off from a hostile exterior. This definition has shaped the way in which we view bodies, and has, in turn, influenced the way in which we define comfort. However, this textbook definition of immunity as 'protection from' is about as faithful to immunity's etymological roots as comfort is to its Latin heritage. In his seminal text, *A Body Worth Defending* (2009), Ed Cohen states that immunity 'now signifies a mode of boundary maintenance that characterises a diverse range of possible actors from bodies to nations to the planet itself'. Immunity, in its biological sense, was coined by Élie Metchnikoff in 1881 as 'form of self-defence' (Cohen 2009: 2). However, there is nothing inherent in the term 'immunity' that coincides with the idea of self-defence, as Cohen shows in his genealogy of the concept. During Roman times, immunity referred to 'privileges and entitlements conferred on individuals or collectives that exempt them from political obligations and responsibilities' (40). The root of the Latin *munis* (immunity stems from *im +munis*) refers to shared duties or services (as in the term municipal). Interweaving the history of the modern, bounded body with the idea of self-defence reaches one of its pivotal moments in the seventeenth century, where, as Cohen demonstrates via his reading of Hobbes, the body becomes property. This allows the body, as well as property, to be governed by law. By placing the Roman citizen outside of the law, Cohen observes that immunity is therefore always a matter of law. With Hobbes's view of body as property, natural law and political law come together, with self-defence relating to defence against loss of property. To own a property and to *have* a body are therefore viewed as fundamental rights.

During the seventeenth century, the body was also seen as a 'discrete, movable mass'. A further pivotal moment in Cohen's genealogy, which comes two centuries later, is the interiorizing of the term *milieu*. An organism lives in balance with its *milieu* or surroundings (Cohen 2009: 188). However, Claude Bernard radically interiorized the old *milieu cosmique extérieur* in the 1850s and 1860s, leading to a new, individualistic conception of health and the body. *Milieu intérieur* was a term which laid the organism open to vivisection in a laboratory environment, and, as Cohen argues, helped to centre the locus of life within the organism, determining a rupture in biological thinking. In opposition to Hippocratic medicine, which advocates that an organism lives in balance with its environment, *milieu intérieur* severs the organism from its environment and renders the body as autonomous, defining the organism's 'internal defence against an actively and relentlessly hostile external word' (130). In addition to being viewed as a discrete entity, and property of the subject, *milieu intérieur* defines the body as singular and in opposition to its surroundings. Noting that 'the theory's meaning and its central metaphor inextricably intertwine precisely because they make the word "make sense", *in a most material fashion*, in new and compelling ways' (35), conceiving of immunity as a form of what Donna Haraway calls 'border war', distinguishes the 'victorious self' from its surroundings (Haraway 1991: 150). Border war has come to define our basic understanding of the term, even though this does not encapsulate the development of the scientific field from the late nineteenth century to the present day. In fact, Bernard even admitted that the freedom and independence that *milieu intérieur* attributes to the organism are an appearance and an illusion (Cohen 2009: 204). Nevertheless, his ideas, Cohen argues, played an important role in Metchnikoff's 'discovery' of immunity as defence, and the wider view of the modern body not only as bounded, but as autonomous and self-maintaining. To the present day, the image of the bounded body continues to define Western subjectivity. It even pervades contemporary neuroscience, performing a central role in Antonio Damasio's theory of consciousness. Damasio writes:

> Life is carried out inside a boundary that defines a body. Life and the life urge exist inside a boundary, the selectively permeable wall that separates the internal environment from the external environment. The idea of the organism revolves around the existence of that boundary. (2000: 137)

Roughly coinciding with Metchnikoff's coining of biological immunity, the development of another modern science was dependent on the idea of a bounded body: Psychanalysis. Freud developed his theory of consciousness founded upon an image of a bounded body – that of a single cell. It is precisely this translation between the biological sciences and psychoanalysis that exemplifies the idea of comfort as an immune system. Consider Georges Teyssot's definition of corporeal comfort:

> One might also recall that, for twentieth-century physiologists and engineers, the notion of comfort is tied to the body's alleged preference for a state untroubled by external disturbance: the zero degree of corporeal excitation. In this view, comfort equals the absence of external stimulation – a sort of sensory weightlessness resembling the 'sensory deprivation' promoted in the United States as a relaxation technique. This state is defined as 'well-being'. (1996: 48)

Teyssot's definition of comfort is both centred on well-being and assumes a bounded state of existence, dependent on an interior defined in opposition to a hostile exterior. Defence against this hostile exterior world comes in the form of anaesthesia; a zone of non-feeling, which can be induced by modern-day techniques, from therapies to engineering practices. One such example is air conditioning, as commented on by Elizabeth Shove, which is a perfect example at how 'scientific' beliefs held about bodies have led to such 'comfort ideals' which are biological impossibilities. The idea that a constant temperature, usually of 22° C, is a standard measure of comfort, has become universally adopted through the development of office air conditioning in the late twentieth century. Here, comfort is based on the 'heat neutrality' model, which states that 'the heat generated by the body is equal to the heat transferred away' (Shove 2003: 29). In this definition, comfort is found in an environment of constant temperature, thought to (albeit incorrectly) replace the body's own thermo-regulation. In this 'synthesized homeostasis' we are said to expend less energy on mere survival and devote our energy to more productive tasks. Avoiding 'thermal stress' defines comfort as a state of corporeal indifference or a 'state of untroubled disturbance' in Teyssot's terms. Teyssot is describing a distinctly Freudian idea here, dependant on the notion of boundedness.

Freud writes:

> Let us imagine living organisms in their simplest possible form as an undifferentiated vesicle of irritable matter; its *surface*, inasmuch as it faces out towards the external word, is thus differentiated by its very position, and serves as the vesicle's receptor organ. (2003: 65)

For Claire Colebrook, Freud's *Beyond the Pleasure Principle* is the 'classic meditation on the image of the bounded body' (2014: 126). Note how Freud views this single cell as a bounded entity, its surface enclosing it from the 'external world'. Not only did ideas of boundedness influence biological immunity as self-defence, they also inspired psychoanalysis, the Ego conceived by Freud as yet another bounded body. The biological metaphors that appear in defining Freud's work are no accident. He began his medical studies at the University of Vienna and in his final year, studied with Ernst Brücke, who among other topics, specialized in cell biology, linguistics, and the nervous system. Although he decided not to pursue a career in biology, Freud continued to publish papers on the nervous system after his time at university (Sulloway 1992: 3–4, 15). In his *Project for a Scientific Psychology* (1895) Freud proposes the principle of 'neuronic inertia', where 'a balance is observed between the quantity of the excitation and the effort required for flight from that stimulus; so that the principle of inertia is not disturbed' (1954: 357). The organism is therefore required to expend an equal amount of energy fighting off stimuli as is exerted upon it. However, Freud grapples with the difference between internal and external stimuli, stating that internal stimuli, such as hunger, respiration, and sexuality cannot be warded off through this process. Instead, they need to be satisfied by the interaction of the organism with the external world. This activity of completion goes against the principle of inertia, 'towards a reduction of its level of tension to zero' (258). In this chapter, Freud appears to incorporate thermodynamic thinking into his model. As Paul Ricoeur observes, his use of the term quantity 'serves to unify under a single concept anything that produces energy' and can refer to internal and external stimuli. It is defined as a summation of excitation homologous to physical energy: it is a current which flows, which '"stores", "fills", or "empties", and "charges" neurones' (Ricoeur 1970: 74). In *Beyond the Pleasure Principle* (1920) we see the development of a theory which states that organic life must avoid negative stimuli, or unpleasure, and seek positive stimuli. Freud dwells on whether such unpleasure results in an increase in quantity or is dependent on quality, inventing a system of perceptual neurones, which translate quantity into quality (Freud 2003: 370). Freud notes that 'the capacity disappears for perceiving sensory qualities which lie, so to speak, in the *indifferent zone* between pleasure and unpleasure' (Freud 1954: 373–4).

Teyssot's definition of comfort as 'the body's alleged preference for a state untroubled by external disturbance' is underpinned by these ideas: the comfort zone is Freud's zone of difference. Furthermore, Freud's stimulus shield which he develops in *Beyond the Pleasure Principle* serves precisely as an immune system, a protective barrier between the Ego and its surroundings. As the outer layer hardens, it decreases in sensitivity to trauma, defending the Ego from what lies outside of it. Although Walter Benjamin already made the connection between the interiority of dwelling and psychoanalysis, describing the self as a house with a cellar in his essay *One Way Street* (Schmiedgen 2009: 149), Teyssot's definition of comfort makes the Freudian underpinnings of our understanding of the term explicit.

Here it is important to note the importance of metaphor referred to by Cohen. As Claude Bernard noted, though the belief existed that the modern body was hard and impenetrable, this did not prevent it also being simultaneously imagined as enclosed by a semi-permeable membrane, as the open system that it is – an idea we find in Freud's *Beyond the Pleasure Principle*. A completely closed body would not be able to sustain life, and neither would a completely open body. Colebrook writes:

> What is required then is a border or membrane that enables communication with an outside, but an outside that is always an outside for this bounded body, and that is managed so as to produce only the alteration and perturbation required for self-maintenance. (2014: 126)

This tension between both protection from and dependence on the outside is precisely the problem Freud grapples with when defining his aforementioned idea of neuronic inertia. According to Colebrook, the modern body, so heavily influenced by Freud, predominantly consists of 'equilibrium, homeostasis and autopoiesis' (2014: 128), bringing us onto a further biological development that has influenced the idea of comfort. Both Nigel Walker and John Bowlby acknowledge that Freud's insistence on avoiding stimuli and his principle of inertia could be read as a precursor to homeostasis, recognizing that the term was not invented for another twenty years (Walker 1956: 61, Bowlby 1969: 23). Interestingly, it is specifically the process of homeostasis that relates comfort to the bounded body in definitions from popular psychology. For example:

> Homeostasis in our psyche is called our 'comfort zones'. This is 'psychological homeostasis' (healingtoolbox.org).

And,

> Often, even when we do try to venture out of it [our comfort zone], we are quickly pulled back in to it. There is a dynamic called 'homeostasis' which is critical to this. Homeostasis has both psychological and physical implications and what it's pointing to is the fundamental and biological drive for equilibrium and stability in a system, (and yes, we are including human beings as systems). In effect, homeostasis helps create and regulate our 'comfort zones'. (www.2130partners.com)

The latter definition above effectively illustrates the belief that the role of homeostasis is for a system to remain in equilibrium, equating comfort to this process. Both of these definitions posit comfort, and more particularly 'the comfort zone' as a state of stability. The 'comfort zone' is also a term used in the realm of business and management, also relating to stability. Comfort, in this sense, is designated as a place of stagnation and limited growth: we are told to get out of the comfort zone. Whereas Cohen shows us that the biomedical world borrows existing beliefs to absorb scientific thinking into the cultural imagination, these definitions of comfort demonstrate that 'poetic borrowing' works both ways, with cultural beliefs drawing on scientific principles to make sense of everyday life, which again, heavily rely on metaphor. Anson Rabinbach's cultural history, *The Human Motor* (1990) is a rich example of the impact of closed system thermodynamics on the way in which people conceived of their bodies. Homeostasis, coined in 1926 by Walter B. Cannon, does not render the body as a closed system, a thermodynamic machine heading towards maximum entropy, or, in Sloterdijk's terms, a 'monadic ego orb'. Rather, homeostasis appreciates that organisms are bounded by semi-permeable membranes, and that to sustain life, exchange must occur between organisms and their environments. The etymology of homeostasis comes from *homoios*, Greek for 'like' and *stasis*, staying the same (*OED* online). However, Canon himself was clear in specifying that the term was not meant to imply a passive body, writing that the organism was 'engaging in free exchange with the outer world' (Canon 1932: 20). He states, 'This word [homeostasis] does not imply something set and immobile, a stagnation. It means a condition – a condition which may vary, but which is relatively constant' (24). It is thus interesting to note the cultural appropriation of this biological term, inscribed in its etymology, demonstrating that the power of the metaphor works in both directions between 'science' and 'culture'.

The form of stasis offered by the body enclosed by a semi-permeable membrane does not signify a fatal entropy-like state (which the term 'the comfort zone' alludes to), but instead relates to self-maintenance. The premise of the bounded body relies on the observation of other bodies, so that self can be distinguished from other. It rests on the recognition that *I*, a bounded body, cannot imagine the interior of another bounded body. (Colebrook 2014: 126). The ideal bounded body, as Colebrook reads from Freud's *Beyond the Pleasure Principle,* must fulfil the following criteria: completeness and self-sufficiency – the former relating to that which it seeks outside of itself. Thus, Colebrook argues that there is always an apparent contradiction relating to the bounded body: for the body to remain enclosed, but to also seek what it needs outside itself to be complete. However, this distinction is only apparent because 'completion is always sought on the organism's terms, always for the sake of the organism' (2014: 127). Both processes come from the point of the organism. Completeness or pleasure, in Freud's terms, is always teamed with risk – the risk of trauma or death, and herein lies the definition of discomfort from the same point of view. If comfort is defined as self-maintenance, discomfort would necessarily be defined as trauma. Comfort, then, defined from the point of view of a homeostatic being, could not be completely static or passive, but would ensure self-maintenance of the organism. Comfort for a bounded body would not be based on an immune system defined by border war, but one defined by border maintenance.

Comfort and situatedness

Freud struggles to move *Beyond the Pleasure Principle*, according to Colebrook, precisely because he cannot move away from the centrality of trauma that is founded upon the image of a bounded body, an enclosed psyche and the organism. Colebrook writes, 'A genuine *beyond* of pleasure and a genuine beyond of the organism and its closed world of meaning would also be beyond trauma, for it could not be regarded as an infraction of the body from outside' (Colebrook 2014: 128). It is therefore necessary to define comfort, as we currently understand it, as an affective phenomenon of the bounded body, from the view of the body.

Donna Haraway uses the term 'situated knowledge' to describe a form of acquiring knowledge from experience, a 'view from a body, always a complex, contradictory, structuring and structured body, versus the view from above, from nowhere, from simplicity' (1988: 589). Boundaries (of bodies) materialize in social interaction. 'Boundaries are drawn by mapping practices; "objects" do not preexist as such. Objects are boundary projects' (595). The idea of vulnerability is fundamental to situated knowledge (1988: 590). This theme is also picked up in Haraway's definition of immunity:

> Immunity can also be conceived in terms of shared specificities; of the semi-permeable self able to engage with others (human and non-human, inner and outer), but always with finite consequences; of situated possibilities and impossibilities of individuation and identification; and of partial fusions and dangers. (1991: 225)

Life as such is constituted as a 'window of vulnerability'. The possibility of the destruction of the organism is also that which enables it to live (Haraway 1991: 230). Like Sloterdijk's theory of spheres, Haraway's immune system advocates intersubjectivity and the active formation and reformation of boundaries and borders. Haraway's idea of situated knowledges thus serves as a useful tool to theorize comfort as an affective process in which boundaries and borders are fluid and changing, and to conceive of immunity consisting of vulnerability as well as boundary maintenance. Rather than lying within the interior of the body or subject, this understanding of immunity allows us to think of comfort in the in-between space between life and death and the interior and exterior.

Non-equilibrium or open system thermodynamics gives us another scientific system whereby to rethink comfort in relation to the bounded body. Whereas viewing the body as such a system does not set it apart from the process of self-maintenance, self-maintenance is here not associated with self-sameness. Non-equilibrium thermodynamics influenced Deleuze's privileging of intensity and critique of extensive processes in *Difference and Repetition* (2010). The concept was introduced by Erwin Schrödinger (1887–1961) at Trinity College Dublin, 1943. Schrödinger looked for ways in which thermodynamics could be applied to biological processes. He specified that all organisms are open systems, taking energy and materials from the environment although they are separated from it by membranes: skin, bark, and shell, and 'trade both matter and energy across their boundaries' (Schneider and Sagan 2005: 26). Recognizing the body as a non-equilibrium system still holds onto the notion of boundedness, but emphasizes the instability of the organism. In other words, bodies give the semblance of stability but they in fact draw sustenance from 'the flux of matter and energy coming to them from the outside world', as Isabelle Stengers and Ilya Prigogine state in *Order Out of Chaos* (1984: 127). While the borders of the body are recognized, they are reconstituted as thresholds and gradients, as active milieus in their own right. Although self-maintenance is a vital component of a non-equilibrium system, such systems draw their stability through their contact with their environment. Completion is rendered active and necessary, rather than being defined by risk and trauma. Therefore, rather than protecting itself from the outside via the creation of a zone of indifference, the organism relies on a zone of intensity, where self-maintenance is not synonymous with stasis. Affect as force is at work here, rather than the affections of an individuated body. While imagining the body as a non-equilibrium system does not do away with the bounded body, it does bring us to immunity's antonym: community – existence in conjunction with the environment. Rather than viewing the body as consisting of Bernard's *milieu intérieur*, milieu refers to the in-between that the term denotes, and well-being becomes a matter of intensity rather than indifference. It is therefore possible to see comfort's etymological roots, to give strength, in a present-day image of bodies and organisms, where strength is not based on an immune system of 'protection from', but gained in the instability encountered in the exchange between the boundaries of bodies and their environments.

The way in which modern bodies and subjects have been imagined through the process of poetic borrowing between the natural sciences, psychoanalysis and culture has led to the understanding of comfort as a form of immune system in our everyday use of the term from business and management to popular psychology. Overlapping definitions of immunity, from those which lean towards 'border war' and hard boundaries, to those that advocate shared sites and fluid fringes, coexist to create overlapping and sometimes contradictory views of occupying a 'comfort zone'. Whatever the nature of these borders and boundaries, the bounded body is essential in our understanding of comfort as an immune system. Open system thermodynamics, Sloterdijk's spheres, and Haraway's definitions utilize ideas around immunity that contain the notion of community, acknowledging the importance of the interaction between inside and outside, positing the comfort zone in the in-between, and moving our understanding of comfort away from the inside towards the edges, boundaries, and borders of life. Whereas Teyssot's definition of comfort interprets Freud's zone of indifference as the comfort zone, open system thermodynamics which influenced Deleuze's idea of intensities provides an alternative understanding of the comfort zone as an active in-between space or milieu.

Note

1 This is translated from the German '*Wohnen als Transitivum – im Begriff des "gowohnten Lebens" z.B*' (footnote 12, p. 966). The adjective *gewohnt* means customary, habitual, familiar, or usual.

References

http://healingtoolbox.org/index.php/k2-stub/item/238-habit-body-as-homeostasis-as-comfort-zones-as-inertia (accessed 18 January 2013).

http://www.2130partners.com/are-you-in-your-comfort-zone-2/ (accessed 18 January 2013).

Bachelard, Gaston (1969), *The Poetics of Space*, translated by Maria Jolas, Boston: Beacon Press.

Bakhtin, Mikhail (1984), *Rabelais and His World*, translated by Helene Iswolsky, Bloomington: Indiana University Press.

Benjamin, Walter (2002), *The Arcades Project*, edited by Rolf Tiedemann, translated by Howard Eiland and Kevin McLaughlin, Cambridge, MA, Belknap: Harvard University Press.

Bowlby, John (1969), *Attachment*, New York: Basic Books.

Cannon, Walter B. (1932), *The Wisdom of the Body*, London: Kegan Paul, Trench, Trubner & CO., Ltd.

Cohen, Ed (2009), *A Body Worth Defending: Immunity, Biopolitics, and the Apotheosis of the Modern Body*, Durham and London: Duke University Press.

Colebrook, Claire (2014), *Death of the Posthuman: Essays on Extinction, Vol. 1*, Michigan: Open Humanities Press.

Crowley, John E. (1999), 'The Sensibility of Comfort', *The American Historical Review*, 104(3): 749–82.

Crowley, John E. (2001), *The Invention of Comfort: Sensibilities and Design in Early Modern Britain and Early America*, Baltimore and London: The John Hopkins University Press.

Damasio, Antonio (2000), *The Feeling of What Happens: Body, Emotion and the Making of Consciousness*, London: Vintage.

Deleuze, Gilles (2010), *Difference and Repetition*, translated by Paul Patton, London: Continuum.

Ferguson, Harvie (2000), *Modernity and Subjectivity: Body, Soul, Spirit*, London: University Press of Virginia.

Freud, Sigmund (1954), 'Project for a Scientific Psychology', in *The Origins of Psychoanalysis*, translated by Eric Mosbacher and James Strachey, 355–445, New York: Basic Books.

Freud, Sigmund (2003), 'Beyond the Pleasure Principle', in *Beyond the Pleasure Principle and Other Writings*, translated by John Reddick, 43–102, London: Penguin.

Gloag, John (1973), *Victorian Comfort: A Social History of Design from 1830–1900*, Newton Abbot: David & Charles.

Grosz, Elizabeth (1994), *Volatile Bodies: Towards a Corporeal Feminism*, Bloomington and Indianapolis: Indiana University Press.

Haraway, Donna (1988), 'Situated Knowledges: The Science Question in Feminism and the Privilege of Partial Perspective', *Feminist Studies*, 14(3): 575–99.

Haraway, Donna J. (1991), 'A Cyborg Manifesto', in *Simians, Cyborgs and Women: the Reinvention of Nature*, 149–81, London: Routledge.

Miller, Daniel (2008), *The Comfort of Things*, Cambridge: Polity Press.

OED online http://oxforddictionaries.com/definition/english/homeostasis?q=homeostasis (accessed 18 January 2013).

Pezeu-Massabuau, Jacques (2012), *A Philosophy of Discomfort*, translated by Vivian Sky Rehberg, London: Reaktion.

Prigogine, Ilya and Stengers, Isabelle (1984), *Order Out of Chaos: Man's New Dialogue with Nature*, London: Heinemann.

Rabinbach, Anson (1990), *The Human Motor: Energy, Fatigue, and the Origins of Modernity*, New York: Basic Books.

Rice, Charles (2007), *The Emergence of the Interior: Architecture, Modernity, Domesticity*, London: Routledge.

Ricoeur, Paul (1970), *Freud and Philosophy: An Essay on Interpretation*, translated by Denis Savage, New Haven and London: Yale University Press.

Schmiedgen, Peter (2009), 'Interiority, Exteriority and Spatial Politics in Benjamin's Cityscapes', in Andrew Benjamin and Charles Rice (eds), *Walter Benjamin and the Architecture of Modernity*, 147–58, Melbourne: Re.press.

Schneider, Eric D. and Sagan, Dorion (2005), *Into The Cool: Energy Flow, Thermodynamics and Life*, Chicago: The University of Chicago Press.

Shove, Elizabeth (2003), *Comfort, Cleanliness and Convenience: The Social Organization of Normality*, Oxford: Berg.

Sloterdijk, Peter (2011), *Spheres Volume 1: Bubbles: Microsphereology*, translated by Wieland Hoban, Cambridge MA and London: The MIT Press.

Sulloway, Frank J. (1992), *Freud, Biologist of the Mind: Beyond the Psychoanalytic Legend*, Cambridge MA: Harvard University Press.

Teyssot, Georges (1996), 'Boredom and Bedroom: The Suppression of the Habitual', in *Assemblage: A Critical Journal of Architecture and Design Culture*, 30, translated by Catherine Seavitt: 44–61.

Valéry, Paul (1977), 'Man and the Sea Shell', in Mathew Jackson (ed.), *Paul Valéry: An Anthology*, 108–35, London: Routledge.

Walker, Nigel (1956), 'Freud and Homeostasis', *The British Journal for the Philosophy of Science*, 7 (25): 61–72.

4

Spacing the interior: The carceral body as heterotopia in contemporary Palestinian Cinema

Robert G. White

> To be on the inside, in this sense, is to speak from, be in, a situation which, paradoxically, you do not control and cannot really be sure of even when you have evolved special languages – sometimes evasive, always idiosyncratic – that only you and others like you can understand … being inside is a privilege that is an affliction, like feeling hemmed in by the house you own.
> Edward Said, *After the Last Sky*. (1986: 52–3)

Edward Said: A thought from the inside

In his 1986 book *After the Last Sky*, Edward Said speaks of Palestinian experience *min al dakhil*, which translates as 'from the interior'. This interior manifests itself in different ways. First, it refers in a tangible geographical sense to Palestinians in Israel, whose status as viewed from those in exile such as Said changed from 'different in a pejorative sense' to 'still different, but privileged' (Said 1986: 51) as the tide of Arab nationalism ebbed, and the status of those *fil-kharij* ('in the exterior') diminished. A second meaning is spatial in a more psychological sense, that is, a psychological and linguistic interiority that is collective, an experience of being on the outside while dwelling in the interior, a space 'always to some extent occupied and interrupted by others – Israelis and Arabs' (ibid.: 53). Said dedicates an entire chapter of *After the Last Sky*, 'Interiors', to exploring this condition of being rendered an outsider within the inside. A conceptual move between architectural, corporeal, and linguistic interiors constantly defamiliarizes any discrete notion of outside and inside.

For Said, thresholds and openings become passages through which to pass, but also spaces that can be breached and entered. 'An open door' he states, 'is necessary for passing between outside and inside, but it is also an avenue used by others to enter' (ibid.: 53). For Elizabeth Grosz (2001) the *outside* is peculiar in that it can only be understood by way of negativity. That is, it is *not* the inside, yet one can never be completely outside, for one is always inside of *something*. Thus the border between outside and inside, interior and exterior is a porous one. Said's interior is not a protected space that shelters one from a hostile exterior, the interior is always already hostile, one is both hemmed in by it and excluded from it. This exclusion was particularly pronounced for Said, since he was unable to enter Israel at the time of writing *After the Last Sky*, and thus witnessed the interior vicariously through Mohr's photographs.

Said's thinking of the 'privileged affliction' that is the experience of *al dakhil* – the interior – will inform my reading of what will be termed the *al dakhil* films – the cinema of the interior – which will be the focus of this chapter. Here bodies are both intrusive and decentred; hemmed in by a certain motor-helplessness, but also encaged by a static, rigid frame. This chapter will attempt to build on Said's problematizing of the interior by proposing a complex topological structure of 'interiority', where 'exterior' and 'interior' pass through and reflect one another. This structure will be examined through a reading of a corporeal network. Corporeality is not to be thought of as a discrete object, but rather a process through which bodies and space articulate one another. Leopold Lambert defines the corporeal as

> the (often violent) relation between the designed environment and bodies. Corporeal politics do not exist in a void of objects, buildings and cities; on the contrary, they operate through the continuous material encounters between living and non-living bodies. (Lambert 2015: 6)

A cinema of the interior

The interior then delimits a specific corporeal network, a relation between space and bodies that complicates and undoes any discrete understanding of inside and outside. This condition of 'interiority' finds its ontological roots in the historical–legal condition of the *present-absentee*. This historical–legal condition emerged in the years after 1948, with the 1950 Law of Absentee Property. This determined Palestinians who left their villages during the 1948 war but found themselves within the new state, as corporeally present within the state, but legally absent from their place of origin. 'Thus', Hillel Cohen (2002) writes, 'the internal refugees who fled to villages in upper Galilee or to Nazareth before these were conquered are defined as absentees though they were in the state and were legal citizens.' This historical–legal status occasions a contemporary trace of ontological displacement visible in the cinematic language of the *al dakhil* directors. The cinematic body of the interior is something both caught within state apparatus and held outside. The condition of present absence marks a different Palestinian body than that seen in the film-making of the occupied territories, particularly that of the West Bank. The radical discordance across time and space in contemporary Palestinian life has been noted by Nasser Abourahme, who asks:

> Can anyone say that Gaza, besieged and on the constant threshold of catastrophe, and Ramallah, where a building boom marks the city's integration into international circuits of exchange, share a temporal order? Or, for that matter, the latter with its refugee camps? Or that these timespaces have not been radically sundered from one another? (Abourahme 2011: 455)

A similar framework can be applied to the question of corporeality. A dynamic, resistant body and its encounter *with*, *through*, and *against* space is a common trope in the cinema of the West Bank, which often articulates an overtly resistant cinematic body in movement (albeit frequently arrested) through space and often against obstacles. A notable example of this is Sharif Waked's video installation *Chic Point: Fashion for Israeli Checkpoints* (2003). This short piece subverts the Israeli soldier's biopolitical gaze, transforming the Agambian bare body[1] into a troublesome, erotic object by juxtaposing real footage of checkpoints with a runway show displaying actor/models. They are provocatively dressed in outfits that reveal their midsections, framed through the soldiers' gaze as always already weaponized. One of these figures – Saleh Bakri – inhabits a markedly different body in Elia Suleiman's *The Time That Remains* (2009), as will be revealed later on in this chapter.

A more recent example that illustrates this resistant body is Hany Abu-Assad's *Omar* (2013), which follows its character through a series of flights, captures, and betrayals. Despite a carceral theme to the film, as Omar is repeatedly captured and released, his is a dynamic body – muscle and sinew – and one which drives a kinetic, albeit claustrophobic mise en scène. Body and surface in this film are in a perpetual state of conflict and negotiation. Leopold Lambert (2013) has highlighted the subversive role of parkour in Gaza and Hanna Baumann (2015) the somewhat casual nature of trespassing in Khaled Jarrar's *Infiltrators* (2012). The separation wall figures in *Omar* not as an insurmountable obstacle. Rather, it is an everyday challenge – the fabric of the city is a haptic (albeit deeply hazardous) cartography for the main character. This focus on the Palestinian body as either a site of resistance or control marks a cinematic body I argue that structures the mise en scène of a number of West Bank films, where power relations are more overt and heightened. Territory here plays a part in this production of bodies in space. The cinematic body of the West Bank is an object of power but – quite literally – an *ob-ject* in the Latin sense of the root, that is, that which is thrown against.[2] This is a body often on a collision course with the physical and juridical obstacles of power.

In contrast to this resistant, dynamic body, the cinematic body of the interior is not marked by kinesis and resistance. Within the interior dwell Palestinian bodies in crisis. These are decadent, pathological bodies, characterized not by kinesis, collision, and resistance, but by stasis, confinement, and enclosure. The work of Elia Suleiman and Kamal Aljafari, which will be the focus of this chapter, articulates a particular ordering of bodies in space expressive of the liminal present absence that characterizes 'interiority'. Indeed, Suleiman's third feature film, *The Time That Remains* (2009) carries the subtitle 'Chronicle of a present-absentee'. Suleiman and Aljafari are two directors whose cinematic language is marked by a relation of bodies to space particular to a subjectivity of being a Palestinian in Israel. Suleiman (from Nazareth) and Aljafari (born in Ramla but raised in Jaffa) share an experience of existence between Europe and what is now Israel. They share official identities as Arab-Israelis but are both Palestinian Christians. Both directors feature in their work, playing a largely observational role.[3] While the directors themselves share a transnational hybridity, their films articulate a corporeal liminality peculiar to *al dakhil*. How their work frames bodies in space is markedly different to the films of the West Bank. The body in the films of these directors is not something that centres or drives narrative and movement as it does in *Omar* (2013), but rather something at the periphery, which both articulates and reflects spatial relations. The body in the work of these two directors is one which is both confined by tight framing but also one which articulates confinement; both a point *in* and producer *of* a network of corporeal space. This thinking of spatial networks with corporeal points stemming from crisis brings us back to Michel Foucault, as we move from the corporeal to the carceral.

Foucault: From bodies in crisis to spaces of deviation

In Foucault's short but dense work on heterotopias, *Of Other Spaces* (1967: 1986), there is an extraordinary and brief movement in which the corporeal is conjured and then displaced by the spatial in his classification of heterotopias. Foucault calls 'crisis heterotopias' places reserved for those in a state of crisis. These are other, or even 'nowhere', places reserved for those in a state of biological flux. The 'crisis' to which Foucault refers is of a fundamentally corporeal nature. Citing puberty, menstruation, reproduction, and ageing, he claims that these crisis heterotopias 'are disappearing today and are being replaced … by what we might call heterotopias of deviation' (Foucault 1986: 25). Interestingly, old age blurs this border, with the retirement home traversing both crisis and deviation 'since in our society where leisure is the rule, idleness is a sort of deviation' (ibid.). Decay and inactivity thus bridge the shift from crisis to deviation. Foucault (ibid.) defines heterotopias of deviation as 'those in which individuals whose behaviour is deviant in relation to the required mean or norm are placed'. Citing examples of rest homes, psychiatric hospitals, and prisons, Foucault's discussion of these 'counter-sites' anticipates his later work on discipline, docility, and panopticonism. Of the latter, Foucault informs us that it 'serves to reform prisoners, but also to treat patients, to instruct schoolchildren, to confine the insane, to supervise workers, to put beggars and idlers to work' (Foucault 1977: 205). Foucault's 'spaces of discipline' which are nascent in *Of Other Spaces* are more fully realized in *Discipline and Punish*, in which a whole taxonomy of disciplinary spaces emerges in the notion of the carceral. Essentially, that which is presupposed to be outside of the norm and exceptional, the figure of the prison, emerges as central to the organization and functioning of power, the emergence of a *carceral network* permeating society. As Foucault writes:

> The frontiers between confinement, judicial punishment and institutions that were already blurred during the classical age, tended to disappear and to constitute a great carceral continuum that diffused penitentiary techniques into the most innocent disciplines, transmitting disciplinary norms into the very heart of the penal system and placing over the slightest illegality, the smallest irregularity, deviation or anomaly, the threat of delinquency. (ibid.: 297)

The (bio)politics of discipline

Foucault alluded to the spatial nature of discipline in his lecture of 17 March 1976. In this lecture, he informs us that

> in the seventeenth and eighteenth centuries, we saw the emergence of techniques of power that were essentially centred on the body, on the individual body. They included all devices that were used to ensure the spatial distribution of individual bodies (their separation, their alignment, their serialization, and their surveillance) and the organization, around those individuals, of a whole field of visibility. (Foucault 2003: 242)

Discipline is thus the distribution of individual bodies in space as a means of ordering, controlling, and subjecting them. In the same lecture, in which Foucault is explicating the meaning and function of what he initially terms 'biopower', he appears to suggest a shift in scale and focus, that is, from the individual to the multitude. To illustrate this, Foucault states:

> So after a first seizure of power over the body in an individualizing mode, we have a second seizure of power that is not individualizing but, if you like, massifying, that is directed not at man-as-body but man-as-species. (ibid.: 243)

Explicating this shift, Foucault further states that biopolitics 'deals with the population, with the population as a political problem' (ibid.). These comments reinforce his initial view as biopolitics being focused on the body-politic rather than the individual body address of discipline. Rather than understanding the relation between the politics of discipline and biopolitics as diachronic, a shift *from* discipline *to* biopower, it makes sense to view them laterally. Arguably, biopolitics in its reach and focus functions on multiple levels. At the macro level it deals with population, 'a set of processes such as the ratio of births to deaths, the rate of reproduction, the fertility of a population and so on' (ibid.).

The individual body is documented, pathologized, and categorized and at the will of a series of mechanisms 'to medicalize the population' (ibid.: 246). Power thus confronts bodies as component parts, but doesn't recognize the individual body. It rather sees the body as a component part of a network – a constitutive element of a wider political problem, the body-politic.

The docile body, a concept Foucault develops in *Discipline and Punish* (1977), is arguably the object of this micro-level operation of biopolitics. 'A body is docile' writes Foucault 'that may be subjected, used, transformed and improved' (Foucault 1977: 136). For Foucault 'discipline sometimes requires enclosure, the specification of a place heterogeneous to all others and closed in upon itself. It is the protected place of disciplinary monotony' (ibid.: 141). While Foucault is ostensibly referring to an enclosed space, his invoking of a space 'closed in upon itself' also isolates the individual body-space, that is, the body as a site of enclosure, heterogeneous to all others. As can be seen from the move between corporeal crisis and spatial deviation in *Of Other Spaces*, at the foundations of heterotopias, or other spaces, lie bodies. While this is implicit in *Of Other Spaces*, the body is more explicitly placed at the origins of heterotopia in a work dating from the same period of Foucault's thought.

Foucault's *Le corps utopique* was first presented as a radio broadcast in 1966, alongside his first broadcast of *Les hététotopies*, which prefaced his better known lecture on heterotopias a year later. *Le corps utopique* initially opposes the body to utopias, alluding to Proust's narrator who awakens to find himself trapped in his body. While utopias are placeless places, the body by contrast is the opposite – the absolute or 'pitiless place' (Foucault 2006: 229). Throughout the essay, Foucault conjures up a series of juxtapositions which displace one another when thinking about the place of the body. It is both the absolute place, articulated against utopias which exceed the corporeal, such as the magical, the soul, and the tomb; yet Foucault also finds in the body 'its own "phantasmagoric resources"' (ibid.: 230). Thus the body is the originator of utopia, a 'utopian actor' (2006) yet always already elsewhere, or perhaps more accurately *nowhere*. 'It has no place' writes Foucault 'but it is from it that all possible places, real or utopian, emerge and radiate' (ibid.: 233). Imagined thus, the body is a conduit, a site in a network through which place is occasioned. It is in the image of the corpse and in sex (death linking the two in French) that the body is here, returning to the corporeal closure of the 'pitiless place' that opens the chapter.

What is striking about *Le corps utopique* is the way in which the otherness of the body, the contradictions, and contested states it holds mark it neither as utopia nor its opposite, but rather heterotopic. Visible and invisible, here and elsewhere and, in a passage that mirrors the fifth principle of heterotopia, both 'open and closed' and 'penetrable and opaque' (ibid.: 231). This marks the body as a site of both isolation and penetration, a system of opening and closing. What emerges from the body through a reading of *Le corps utopique* (1966), *Of Other Spaces* (1967), and *Discipline and Punish* (1977) is the body as heterotopic site, a system of opening and closing through which power functions spatially. This is not a body *in* space; rather, it is a body which reflects and occasions space – a body *as* space.

Jean-Luc Nancy – the open space of the body

Jean-Luc Nancy radically inverts a geometric thinking of bodies in space and postulates a thinking of the corporeal not as a relation, but closer to Merleau-Ponty's (1968) thinking of an intercorporeal ontology of flesh, where living and non-living bodies intertwine. Nancy claims that 'the world of bodies … is a world where bodies initially articulate space. The world is spacing, a tension of place, where bodies are not in space, but space in bodies' (Nancy 2008: 27). Nancy's *Corpus* (2008) displaces the body in a strikingly similar movement to that of Foucault forty years earlier. For Nancy, the body is disruptive to thought, in a manner akin to Foucault's utopian body. The body, as thought by Nancy, appears not to properly inhabit either matter or discourse. It is rather a limit concept, touching *on* and moving *between* stable, discrete notions of mind, body, and soul. For Nancy:

> *Bodies don't take place in discourse or in matter*. They don't inhabit 'mind' or 'body'. They take place at the limit, *qua limit*: limit–external border, the fracture and intersection of anything foreign in a continuum of sense, a continuum of matter. An opening, discreteness. (Nancy 2008: 17; emphasis in original)

The body is thus between matter and meaning, subject and object, inside and outside. It is an opening but also an enclosure. It is both interior and exterior. At these borders we find Said's interior, a privilege that is also an affliction, and a Foucauldian body, a body always already nowhere and elsewhere. There is also a founding alterity to Nancy and Foucault's body. Both *Corpus* and *Le corps utopique* open with the burden and otherness of the body, with Foucault lamenting the impossibility of escaping it, and Nancy (2008: 7) highlighting its 'prison-wall thickness'.

While giving primacy to thinking the body as a 'maker' of space, Nancy does mark a distinction between what we might think as Euclidian space and the space that the body enacts, for he sees the body as *making room* for existence, arguing that bodies are 'more properly spacious than spatial, what could also be called a *place*' (ibid.: 15).

What Foucault and Nancy share in their thinking of the body is its production of space, more properly as sites through which space and place can emerge. For the former, it is 'where paths and spaces come to meet' (Foucault 2006: 233) and the latter a spacing through which 'the essence of existence is to be without any essence' (Nancy 2008: 15). What can be taken from Nancy and Foucault's radical inversion of bodies in space, particularly when read through Foucault's later work on discipline, is the emergence of a biopolitical network of corporeality with which to read the relation between bodies and space in the work of Suleiman and Aljafari. Nancy's (2008) thinking of the body as 'existence without essence' meant to liberate the body from its traditions can, through the lens of biopolitics, be rearticulated as existence without being – displaced and disjointed bodies outside of time and place, but nonetheless confined to interiority. This is not the body as open space – that which spaces space – but rather the body as enclosure, that which confines the space it occupies. If there is a plasticity at work here, it is a plasticity of confinement which articulates space through bodies. What emerges in the cinema of the interior is not merely a corporeal network, but bodies themselves as sites in a carceral network, a network of body spaces that are both confined and articulate confinement. When speaking of the tension between utility and subjection in the docile body, Foucault writes:

> (discipline) dissociates power from the body; on the one hand, it turns it into an 'aptitude', a 'capacity', which it seeks to increase; on the other hand, it reverses the course of the energy, the power that might result from it, and turns it into a relation of strict subjection. (Foucault 1977: 138)

However, the biopolitical, docile body-space we see in the cinema of the interior is not one of capacity, but rather one of inertia. What we witness is the pathological body. This is the body as subjection – subjection as enclosure. Bodies in the work of Aljafari and Suleiman are not the explicit objects of discipline and resistance that you see in the more overtly biopolitical space of the West Bank; rather, they are covert producers and products of a carceral space. Docility and confinement are not so much centred *on* the body, as *in* and *from* the body.

Elia Suleiman: Interior bodies in crisis

Elia Suleiman's *The Time That Remains* (2009) is the third film in what might be termed the 'present-absentee' trilogy. All three films focus on bodies of the interior, bodies of stasis, ageing, and decadence, particularly in the Nazareth scenes. *The Time That Remains* (2009) can be read as a film about the body's trajectory from resistance to docility. The film marks something of a break with Suleiman's earlier work in that it is ostensibly more rooted in time and place.[4] The film is largely chronological and has four distinct temporal markers: 1948, 1970, 1980, and the present day. Historical events (the Nakba, the death of Nasser, the Land Day Protests) mark the transitions. Whereas in his two previous films the site has been a dislocated present, and Suleiman (or his cinematic alter-ego 'E.S.') has been at the centre of the films, in *The Time That Remains*, E. S. is a more peripheral figure. A large part of the film is dedicated to his family in general, his father Fuad, in particular. Or perhaps more accurately, the body of his father. For this is primarily a film about bodies. The body of the father. The dynamic body becoming the body of stasis. The medicalized, docile body of entrapment. As the film unfolds, its kineticism dissipates and its framing becomes more static. We see the emergence of pathological bodies, often both trapped in these bodies and confined to domestic or clinical interiors. The enclosure of the bodies of Suleiman's parents is mirrored by the enclosure of the spaces that confine them.

A key scene appears towards the middle of the film. The scene takes place outside a pharmacy, as E. S. collects the prescription issued for his father. As we see E. S. enter the pharmacy in a long shot, the scene cuts to the car, where his father puts a cassette into the player. The song is 'Ana Albi Dalili' by Laila Mourad, an Egyptian singer. This is the same song Fuad hears on a stolen record player when he is hiding from Israeli soldiers in Nazareth in 1948. As the song plays in the car, Fuad nostalgically nods along, as E. S. stops and turns to watch his father drifting into sleep, his athletic, resistant body of youth replaced by a docile, decadent hospitalized body. Similarly, we see Suleiman's mother largely confined to the Nazareth apartment, an almost motionless figure whose movements are made after dark, as she covertly eats ice cream at night, which leads to her being chastised by her diabetes nurse as her health deteriorates.

The unhealthy, docile bodies we see evolve throughout *The Time That Remains* (2009) are more explicit in Suleiman's previous film, *Divine Intervention* (2002). There are a number of hospital scenes, reflecting the father's sickness we see developing in *The Time That Remains* (2009). In the first of these scenes, shortly after we see the father collapse in the kitchen after taking his morning coffee, we see a ward of ageing Palestinian men, the latest casualty being wheeled in.

This first hospital scene begins with a long shot of the exterior of the hospital, with the non-diegetic beeping of monitors as cars pass along the road below. As the scene cuts to the hospital ward, we see E. S. in the far corner looking out across Nazareth. In the ward we see two men in the beds, the furthest being E. S.' father in the background. In the foreground, the new patient arrives to complete the mise en scène with a striking symmetry. The two foremost characters lie supine on their trolleys, their protruding stomachs rising and falling in synchronized patterns. Despite the clinical compartmentalization of the hospital trolleys, the foregrounding of these bodies creates a landscape of corporeal crisis, a rearticulation of home/sickness, a sickness not as longing for a home elsewhere, but a malady dwelling in these bodies trapped in the interior. These two men form a mirror image of ill health, their profiles projecting a pathetic image of the sick body as incarceration.

The biopolitical, docile body of the interior reduces the body itself to a space of confinement. The body becomes a carceral space. Characters become trapped in unhealthy, immobile bodies, which produce an ever-constricting, tableau-vivant-like mise en scène. Indeed, the figure of the tableau vivant itself has been theorized as a disciplinary space by Foucault in *Discipline and Punish*. When speaking of the spatial ordering which discipline occasions, Foucault writes:

> The first of the great operations of discipline is, therefore, the constitution of '*tableaux vivants*', which transform the confused, useless or dangerous

multitudes into ordered multiplicities. (Foucault 1977: 148; emphasis in original)

While speaking of physical space, Foucault's language evokes the arrangement of bodies and objects in the cinematic frame. In doing so, the cinematic *tableau vivant* is reconfigured as a disciplinary space, a space that subjugates bodies to separation, ordering, and cataloguing. There is essentially an enframing at work in the static frames and compartmentalization of bodies in the hospital scenes of *Divine Intervention* (2003).

The interiority of the outside: Surfaces, walls and carceral bodies in the work of Kamal Aljafari

A similar enframing and compartmentalization of bodies can be seen in the mise en scène of Kamal Aljafari's work. Aljafari's *The Roof* (2006) is a film articulating Edward Said's 'interior'. The film is semi-autobiographical, as it concerns the director (who is based in Germany) visiting his mother's home in Ramle and his father's home in Jaffa. The film opens with a flattened, compressed shot of a rain-blurred window. We hear the director off-screen, describing his experience in a prison for six and a half months during the first intifada. Few details are given; we hear fragments of his experience, as the face of his sister fills the frame, listening and questioning. The interior in this case is a non-place, perhaps a bar, café, or hotel lobby, the mise en scène is opaque.

The description of the prison is matched by a formal claustrophobia. The shallow focus and tight framing bring his interlocutor's profile looming into the foreground while the background is abstracted – only raindrops streaming down a window – adding to the any-place-whateverness of the location. The off-screen, disembodied voice decentres this subject from frame while retaining a psychological interiority, off-screen. This is the de-cantered 'inside' of Said (1986), speaking from a situation outside of one's control and certainty. The scene also uses a dislocating sense of irony that occurs intermittently. While narrating his experience of prison, the friends he made and lost touch with, Aljafari explains that he fell out of touch with his friend Nabieh 'because I didn't know what to say to him – I was free and he was in prison'. These words ring hollow, as both characters are held captive by the static flattened frame and shallow-focus photography.

The composition of this opening scene anticipates the cinematic language of the film. Static, tableaux-like frames pass along (or linger at) walls, windows, and unfinished roofs. Interiority becomes a mode of seeing in these scenes. Whether the scene focuses on roofs, the open sea of the port of Jaffa or literal interiors, a spatial language of enclosures and enclaves is employed, rendering the interior here not as a site of privilege, but rather one of being-hemmed-in.

Port of Memory (2009), Aljafari's second feature film is, like *The Roof* before it, a film about carceral bodies trapped in the interior. The film takes place in Jaffa, the port of the title and the narrative concerns the loss of the deeds to the main character Salim's house. The threat of eviction hangs over the characters. Like *The Roof*, its predecessor, the character of the film is arguably the architecture itself, as the human characters are often dwarfed by the interiors and the cinematic language regularly forecloses anthropocentrism. Often at the edges of frames or shot from windows or doorways, the characters are dispersed and dislocated. Exterior shots have a similar formal composition to those in *The Roof* with long shots highlighting the dilapidated architecture of Jaffa, with characters occasionally passing into and out of frame. The interior scenes range between long, architectural tableaux and frequent close-ups of specific elements of the mise en scène. It is important to note at this stage that the original Arabic title for *The Roof* is *al Sateh*, which gives an ambiguity lost in translation. Alongside its usage here, this term can also convey a more general sense of flatness, or surface. It is precisely this which provides a conceptual link between the two films. The surface and the wall are the focus in *Port of Memory*, which displays a spatial language speaking only of interiority. While there are ostensible 'exterior' shots, these outside spaces are really only thresholds into further interiority.

A recurring scene in the film takes place in a non-descript café in Jaffa where the three characters or bodies, as we are never given any biographical detail, sit motionless in a café, at the edges of the frame in a static long shot. The tableau-like geometry of the scene is emphasized by the grid-like bars on the windows. The only movement that cuts through this still image is the pathological gesture of one of the characters holding a burning coal from the fire inches from his neck, before placing it down again. These bodies, docile and inactive are compartmentalized by this spatial arrangement which renders them objects of the mise en scène, the lines between living and non-living bodies blurred – these are bodies *as* space. When writing of Clair Denis's framing of bodies in space in *Beau travail* (1999), Douglas Morrey describes the legionnaires as

> an illustration of Nancy's concept of the spacing of space: an attempt to make sense of physical space through bodily action, an active – even aggressive – inhabiting of space through a disciplined occupation of the body. (Morrey 2008: 13)

However, what we see in the café scenes of *Port of Memory* is carceral space occupying bodies. There is no real hierarchy here between the living space of bodies and the architectural space that frames them. These are bodies as landscape, a network of living and non-living bodies articulating confinement.

The opening scene of *Port of Memory* frames the interiority of the spatiality of the film, much in the way *The Roof* did in its opening scene. The scene opens with a slow, quiet panning shot of the exterior of a house, the only audible sound that of morning birdsong. The scene is shot at twilight, which casts a blue hue over the stone walls. The camera tracks left along walls, revealing the scars of neglect, a broken balcony, bricked up windows and a faded frieze. The jagged metal struts jut out like spikes, giving the house a slightly menacing air. A once grand home, in this tightly framed, cool blue tracking shot is rendered as a looming, solid wall.

This scene sets up what can be called a language of confinement that runs through *Port of Memory*, much like it did in *The Roof*. Exterior walls play the role of hemming in the characters, who are often dwarfed by them. Another scene which is repeated follows Salim as he brings food to a character named Samir (who remains off-screen and unseen) in a seemingly abandoned house. The scene is striking in that it is a putative external long shot, but feels claustrophobically interior. The camera remains static as Salim wanders into, and then out of, the shot. Both walls tower above him and the scene feels prison-like. When combined with the diegetic exterior sounds of constant construction, as Tel Aviv bears down on what remains of Jaffa, the sense of the interior as a space of confinement is oppressive.

This act of enframing immobile bodies that we witness in Aljafari and Suleiman's interior produces bodies as enclosure; bodies as space, not the open space of Nancy's corporeality, but rather a closed disciplinary space which allows a disciplinary network, or perhaps more accurately, a *circuit* to function. Indeed, in his lecture at the Collége de France of 17 March 1976, in which he sketches out his theory of biopower, Foucault references the figure of the circuit through which biopower flows. In a passage in which he speaks of its reach over the medicalization of the body, and the threat of infirmities and accidents, he warns that 'they have similar effects in that they incapacitate individuals, put them out of the circuit or neutralize them' (Foucault 2003: 244). This 'circuit', to which Foucault (ibid.) also refers as a 'field of capacity', is explicitly spatial in its imagery – that of bodies as points in a circuit – through which, when complete (or closed) biopower can function. Any openness or removal of these bodies thus breaks the circuit. Opening is thus an act of resistance, in that it interrupts the flow of biopower. It is just such a break in the link that occurs in *Port of Memory*, opening a space through which a resistant subjectivity can emerge in the interior.

Pathways to resistance: Heterotopias of illusion and topological perforation

Towards the end of *Of Other Spaces*, Foucault appears to allude to the critical function of heterotopias, as truly heterogeneous sites whose role it is to expose the illusory nature of real spaces. That is to expose as artifice that which has been naturalized, through juxtaposition. When speaking of 'a function in relation to all the space that remains' (Foucault 1986: 27), he writes of creating 'a space of illusion that exposes every real space, all the sites inside of which human life is partitioned, as still more illusory' (ibid.) Foucault cites the example of the brothel as a site of escape which highlights the illusory nature of a society supposedly founded on monogamy and fidelity. This sixth principle is where the heterotopia seems to become most explicitly a space of critique, a 'counter-site' (ibid.: 24) of contestation. This is primarily a space from the outside which punctuates the artifice of the inside, that is, the putative 'real' space.

While *Port of Memory* (2009) is primarily a film about the interior, where bodies occasion carceral space and enact an inert form of disciplinary power, there is a key scene which breaks this circuit of power, a scene of radical openness in which the present-absentee of the interior acts back from the margin onto the centre that has excluded him. The film occasionally foregrounds the liminal and spectral, taking the unwitting Palestinian extras in the *Delta Force*[5] – which an unnamed character is seen watching – and recognizing their disruptive, spectral present absence as troubling to a fictional unity of an Israel which excludes its 'interior' Palestinians. In the key scene, which takes place towards the end of the film, a clip is shown of Salim, walking the streets of Jaffa, towards the old port. The quality of the image is somewhat bleached, a notable difference from the cool but crisp palette of the rest of the film. The reason for this is revealed, as this is another Jaffa, a site of illusion and counter history. The scene begins with a graphic match, as Salim (played by Aljafari's uncle) is inserted into shots from which the scenes were taken, an Israeli film *Kazablan* (1973). This film manages to layer a fictional cinematic occupation on top of the factual occupation of Jaffa during this period, as the films tells a narrative of oppressed Mizrahi Jews living in Jaffa, and the scene in question is the sung lamentation of Ashkenazi oppression, a narrative which, as Aljafari states, 'completely elides not only Jaffa's Palestinian history, but also its remaining Palestinians, enacting a virtual, cinematic emptying of the city' (Himada 2010).

As the character wanders the crumbling architecture of an abandoned Jaffa, he sings the lyrics to Yesh Makom, his song to his former life in Morocco:

> There is a place beyond the sea,
>
> Where the sun shines over the market, the street and the port,
>
> Home beyond the sea…

Salim's spectral appearance troubles this scene, briefly haunting the frame from the edges and fracturing and undermining the fictional narrative of the scene's Mizrahi lamentation. By reappropriating a Hebrew song of loss and longing the scene counters a hegemonic space, the national territorial, by highlighting the fictional unity of the centre. The fictitious elision of the Palestinian, or the *Arab al dakhil*, is countered and perforated by the claim that there have always been *peoples* in Israel–Palestine, never simply *a people*. The film both enacts a rearticulation of space and decolonizing of the image, a form aesthetic resistance to what Aljafari himself has referred to as the 'cinematic occupation' (Himada 2010) of Jaffa.[6]

The interior here is ruptured from the outside, as an exclusionary space which contains but conceals the Palestinian as present-absentee. Said's figure of the outsider here laterally perforates a fictional inside as just that – an untenable fiction. The national-territorial myth of the projection of emptiness, reinforced by the cinematic emptiness in Israeli films, such as *Kazablan* (1973), is punctuated by the spectral, disruptive presence of the figure of Salim. This scene occasions a new space, a space of possibility, with its use of the virtual to punctuate the fictitious.

The figure of the present-absentee, a legal citizen but an ontological internal refugee, always already displaced, here sunders the homogeneous space of the nation-state depicted in *Kazablan* and offers an image of a resistant subjectivity, not the subjection of the carceral body heterotopia seen in much of the interior of Suleiman and Aljafari's work, but rather a claim to be recognized as always already present in this space. By highlighting the illusion of the absent Palestinian in Israel, the scene acts not only as heterotopia of illusion, but also has a more radical function.

In his 1993[7] essay 'Beyond Human Rights', Giorgio Agamben, drawing on Arendt, calls for a separation of the concept of people from that of nation, marking a new space, a space which he claims would

> coincide neither with any of the homogenous national territories nor with their *topographical* sum, but would rather act on them by articulating and perforating them *topologically* … where exterior and interior in-determine each other. (Agamben 2000: 25)

The essay ends with Agamben's reference to the expulsion of around 400 Palestinians to the border with Jordan in December 1992, this border space at that very moment, says Agamben, beginning to 'act back onto the territory of the state of Israel by perforating it and altering it in such a way that the image of the snowy mountain has become more internal to it than any other region of Eretz Israel' (ibid.: 26). This topological perforation of the space of the state is precisely the manoeuvre enacted in the scene. Israel's Palestinians, its outsiders within the inside, whose bodies are docile and through whom a disciplinary power is dispersed, here act back on the cinematic space from which they have been long excluded. The scene enacts a minoritarian[8] resistance by both revealing and simultaneously disrupting the spatial politics of Jaffa and reappropriating Hebrew song of oppression and loss, thus deteritorializing the majoritarian language as an act of aesthetic resistance.

Conclusion: The digital as minoritarian resistance in the cinema of the interior

The resistant politics of the cinema of the interior arguably dwells in its aesthetic use of the digital to perform acts of minoritarian resistance, which open up spaces to resist totalization and retain a capacity to critique and to question.

Such a strategy can be seen particularly in Aljafari's use of the digital in *Port of Memory* to deterritorialize the power of the majoritarian. Yet Elia Suleiman's *Divine Intervention* (2003) also uses such an aesthetic strategy to critique overcoded symbols. The film opens with a long shot of a hill in Nazareth. A panicked figure in a Father Christmas suit zigzags up the hill, as presents tumble from his sack. A group of young boys follow his wayward movements as they pursue him. The scene continues until we see a close-up of the out of breath Father Christmas, and the camera pans back to reveal a knife in his stomach. The opening scene prefigures the defamiliarization of time and place and non-linear structure that characterize the film. Further, it pushes the limits of representation, presenting a collection of images at once recognizable

(Christmas, Nazareth) and simultaneously violently out of place (the chase and stabbing). A number of scenes in the film deconstruct majoritarian symbols of fixed identity and power structures. The face of Arafat, the *kuffiyeh,* and *fida'I* are all symbols which have become overcoded in the structure of a 'Palestinian Identity'. These symbols are held up as problematic representations of identity and deconstructed in often absurd ways, such as the scene where Suleiman's cinematic alter-ego, 'E.S.', releases a balloon carrying the wrinkled, distorted face of Arafat, which floats above the Temple Mount, entrancing Israeli soldiers below. These acts of non-representation and deconstruction of fixed notions of identity are perhaps the true spaces of critique, heterotopias of illusion that highlight the illusory, overdetermined real space of Israel–Palestine.[9]

Elia Suleiman (2010: 4) has spoken of his own resistant subjectivity in such ways, when asked of the Palestinian state to come, stating 'I will be fighting until the flag has risen. But then I will be fighting to lower that flag again, because I don't believe – nor in flags, nor in linear identities.' Perhaps this is the space of critique that the cinema of the interior aspires to, what Agamben (1993) refers to as 'extraterritoriality'. To resist is to stand with but retain the capacity for critique; that is, to always remain on the outside.

Notes

1 Agamben marks a distinction between the camp and prison, stating 'the camp is topologically different from a simple space of confinement' (Agamben 1998: 20). This chapter marks a distinction between the non-place of the 'camp' and Foucauldian 'disciplinary space' of the state, a specific relation of power to citizen in Israel, arguably the biopolitical nation-state par excellence. This relation is distinct from the Agambian 'non-citizen'.

2 The Latin term *obicere* being constructed from *ob* (in the way of) and *jacere* (to throw).

3 The framing, wanderings, and propensity to see rather than act puts Suleiman and Aljafari in a lineage of Bazin and Deleueze's neo-realism. What Deleueze refers to as 'any spaces whatever'– 'deserted but inhabited, disused warehouses, cities in the course

of demolition or reconstruction' (Deleuze 1989: xi) – are particularly apt to describe the spaces of Aljafari's Jaffa.

4 Suleiman (2000: 99) stated in an interview that he was not happy with scenes in his *Divine Intervention* (2002) 'that are anchored in temporal reality, linked to a specific period and place that will not last'.

5 At the 2012 Conference on the Palestinian Image in London, Aljafari, a Jaffa native, highlighted the presence of people he knew from growing up in Jaffa, who had been inadvertently caught at the edges of the frame in the filming of a number of Hollywood action films, including *The Delta Force* (1986) in Jaffa in the 1980s.

6 This occupation of the image has a long history in Jaffa. Menahem Golan was producer and director of a number of Israeli and Israeli-American musicals and action films, including *Kazablan* (1974) and *The Delta Force* (1986) – using Jaffa as the stage for a tale of Mizrahi oppression in the former, and a stand-in for Beirut in the latter, as Chuck Norris battles Lebanese 'terrorists' portrayed by Robert Forster and a number of Israeli actors. Golan was a native of Jaffa, although as Aljafari (2010) has noted, his Jaffa is conspicuous in its absence of Palestinians.

7 Published in Giorgio Agamben (2000) 'Means without End. Notes on Politics' in *Theory Out of Bounds, Vol. 20.* Minneapolis/London: University of Minnesota Press.

8 A minoritarian cinema is not interested in representation of a people, but as Deleuze argues, the creation of a people. There is not 'a people' per se to be represented, but rather a multiplicity. As Deleuze states, there are 'always several peoples, an infinity of peoples, who remained to be united, or should not be united ... because the people exist only in the condition of minority, which is why they are missing' (Deleuze 1989: 220).

9 The space of Israel/Palestine, as Edward Said (1986), Lina Khatib (2006), and Eyal Weizman (2007) have all recognized, is always already overdetermined as a site of historical claim, archaeological and ideological conflict, and the epicentre of three monotheistic religions.

References

Abourahme, N. (2011) 'Spatial Collisions and Discordant Temporalities: Everyday Life between Camp and Checkpoint', *International Journal of Urban and Regional Research*, 35(2): 453–461, *Blackwell Publishing.*

Agamben, G. (1998), *Homo Sacer: Sovereign Power and Bare Life*, translated by Daniel Heller-Roazen, Stanford: Stanford University Press.

Agamben, G. (2000), 'Beyond Human Rights', in G. Agamben (ed.), *Means Without Ends: Notes on Politics*, trans. Vincenzo Binetti and Cesare Cesarino, 15–26, Minneapolis and London: University of Minnesota Press.

Baumann, H. (2015), 'Bodies on the Line: Somatic Risks and Psychogeographies in Urban Exploration and Palestinian "Infiltration"', in L. Lambert (ed.), *The Funambulist Papers, Volume 2*, 1st edn, 27, New York: Punctum Books.
Cohen, H. (2002), 'The Internal Refugees in the State of Israel: Israeli Citizens, Palestinian Refugees', *Palestine-Israel Journal*, 9(2), *Middle East Publications*. Available at: http://www.pij.org/details.php?id=159
Deleuze, G. (1989), *Cinema 2*, translated by Hugh Tomlinson and Robert Galeta, London: The Athlone Press.
Foucault, M. (1977), *Discipline and Punish*, translated by Alan Sheridan, London: Penguin Books.
Foucault, M. (1986), 'Of Other Spaces', *Diacritics*, 16(1): 22–7.
Foucault, M. (2003), *Society Must Be Defended*, New York: Picador.
Foucault, M. (2006), 'Utopian Body', in C. Jones (ed.), *Sensorium Embodied Experience, Technology, and Contemporary Art*, 1st edn, 229–34, Cambridge, MA: MIT Press.
Grosz, E. (2001), *Architecture from the Outside*, Cambridge, MA: The MIT Press.
Haider, S. (2010), 'A Different Kind of Occupation: an Interview with Elia Suleiman', *The Electronic Intifada*, 01 February 2010 [Online]. Available at https://electronicintifada.net/content/different-kind-occupation-interview-elia-suleiman/8654 (accessed 02 April 2016).
Himada, N. (2010), 'This Place They Dried from The Sea: An Interview with Kamal Aljafari', *Montreal Serai*, 28 November 2010 [Online]. Available at: http://montrealserai.com/2010/09/28/this-place-they-dried-from-the-sea-an-interview-with-kamal-aljafari/ (accessed 23 July 2013).
Khatib, L. (2006), *Filming the modern Middle East*, London: I. B. Tauris.
Lambert, L. (2013), *The Funambulist Pamphlets, Vol. 06_Palestine*, New York: Punctum Books.
Lambert, L. (2015). *The Funambulist Papers, Volume 2, 1st ed.* New York: Punctum Books
Merleau-Ponty, M. (1968), *The Visible and the Invisible*, translated by Alphonso Lingis, Evanston: Northwestern University Press.
Morrey, D. (2008), 'Open Wounds: Body and Image in Jean-Luc Nancy and Claire Denis', *Film-Philosophy*, 12(1): 10–30. Available at: http://www.film-philosophy.com/2008v12n1/morrey2.pdf
Nancy, J. (2008), *Corpus*, New York: Fordham University Press.
Said, E. and Mohr, J. (1986), *After the Last Sky*, London: Vintage.
Shohat, E. (2010), *Israeli Cinema*, 2nd edn, London: I. B. Tauris.
Weizman, E. (2007), *Hollow Land*, London: Verso Books.

Filmography

Chic Point: Fashion for Israeli Checkpoints (Sharif Waked, 2003: Israel, Palestine)
Chronicle of a Disappearance (Elia Suleiman, 1996, Israel, Palestine, USA)
Divine Intervention (Elia Suleiman, 2002, France, Germany, Morocco, the Netherlands, USA)

Kazablan (Menahem Golan, 1973, Israel)
Omar (Hany Abu-Assad, 2013, Palestine)
Port of Memory (Kamal Aljafari, 2009, Germany, France, UAE)
The Roof (Kamal Aljafari, 2006: Germany)
The Time That Remains (Elia Suleiman, 2009, France, Belgium, Italy, United Arab Republic, Great Britain)

5

The politics of the hidden space: Georges Bataille and non-knowledge in the era of transparency

David Hancock

We live in an era of transparency. Openness and visibility are taken as absolute goods and a cornerstone of the democratic system. The opposites, secrecy, invisibility, and hiddenness are received with suspicion. The doctrine of open government holds that the public has the right to access state documents and proceedings and to monitor the processes of the state, and Freedom of Information Acts now gives citizens the right to access information held by public bodies. Surveillance in urban areas is now ubiquitous. Through the revolution in digital technology, vast amounts of personal information are now willingly shared and even more is collected. Within this context the citizen is also subject to the regime of transparency. Security services have taken advantage of technological change and, enabled by law, now routinely collect and monitor electronic communications. This monitoring is done in the name of national security and through a logic of 'if you have nothing to hide, you have nothing to worry about'. Being open in this sense is taken as a sign of one's respectability while closing one's self off from the public gaze is to attract curiosity and suspicion. Transparency, furthermore, is taken as a prerequisite of a market society that functions on equality of access to information. Asymmetric information is understood to distort market mechanisms to which transparency is the answer to make a free market efficient. Data, through which the minutiae of life is coded, has become a basic commodity of the platforms upon which twenty-first-century capitalism sit, on what is fast becoming an unassailable position over social life.

This chapter considers the heterotopian promise of the hidden space as something other to the neoliberal regime of transparency. Through an analysis of the ontology of the secret, and what the philosopher Georges Bataille called the experience of non-knowledge, this chapter proposes the habitation of hidden spaces as part of the acceptance of the paradigm of non-knowledge. The experience of non-knowledge is described by Bataille as viscerally uncomfortable, and I will argue that it through occupying this space that we can remove ourselves from neoliberal reason, and the regime of digital transparency. This might not mean complete removal of oneself from the visible world but an occupation of a space that is both seen and unseen. Accepting a fundamental non-knowledge through such an occupation of digital space is, I argue, an attempt to move beyond the logic of modernity, the attendant logic of transparency and the law of economic reason.

Transparency in the digital space

The concern of some over surveillance is more than matched by the curious ambivalence and acceptance of the many, particularly in the digital space. For years millions of people have freely given the details of their weekly shopping to the supermarkets. Shopping centres like London's Westfield now offer customers the opportunity to download apps that collate preferences so as to micro-target promotions while they shop. Since the advent of social media millions more freely share information about themselves to billions of others and this data is the core of the internet's social media and search economy. The individual is tracked and that data is then sold to advertisers, making the user a commodity.[1] In the eyes of the tech companies these people understand that the internet economy is based on the gathering and marketization of freely given data. What is proclaimed is a service that is then exchanged for information and sold to a third party. The knowing citizen is freely entering a rational exchange with a company seeking to offer the individual choice. In reality, many people give little thought to their privacy when they divulge their information. They are however happy to have the optimized search that Google offers and like the fact that mobile phone location information pinpoints the individual on a map and shows them where the nearest supermarket is. The internet economy has been built on this transparency and though this is seen as a great mistake by some (Lannier 2013), the convenience and efficiency that is generated through this sharing of information has often been a boon. For example, though I am personally concerned about my online privacy I usually use Google as a search engine rather DuckDuckGo (which doesn't collect and marketize my search history). This is because the data that Google has collected about me makes their search engine better. Google maps collates the GPS information of all users of the app to give drivers accurate information of the speed of traffic and to the quickest route. The more information that is shared, the better the service becomes. The same ambivalence can be detected in the case of government surveillance. When the Snowden revelations about NSA spying came to the public's attention there were two things that jumped out about the reaction. First, for some people there was a complete lack of surprise. These people had assumed that the security services monitored communications and were already uneasy. All that Snowden did was to confirm what was already assumed. What Snowden revealed was not a surprise but a confirmation and a very interesting description of the process and scale. But for many others there was not the same sense of unease. These examples point to the ambivalent relationship that many people now have with transparency and their formally private world (see Coll 2012), they are aware of the problems but in many cases are welcoming of the pay-off.

We need to be clear about what data transparency is. We are not thinking about a file with names on in an anonymous building. Data is collected on all, not only the suspicious. It is therefore curiously democratic and this may help to explain the generalized ambivalence towards it. Data is not present-to-mind knowledge and a person often only appears to another individual at the population level. The micro-targeted adverts at the Westfield shopping centre in West London do not pass through an individual but a machine; the data is processed through an algorithm that then produces an output that an individual may or may not encounter. Within this context the citizen has become, as Claire Birchall has noted, a data subject, a mere part of a data set (Birchall 2015). The question of secrecy, as I will come to later, is a social mechanism that relies upon the interaction between subject and subject so it may be felt that there is no breach of secrecy taking place here, or at least not one that is viscerally noticeable in the same sense as a Stasi file. Indeed, the power of the panopticon lies in the assumption that the prisoner makes regarding the presence of surveillance, despite not seeing it. This is not the same with the digital which is opposite in that it is ubiquitous yet often unseen. However, and this is a point recognized by Howard Rheingold, a key participant in the WELL online community in the early years of the internet, who noted that the mediatization of the panopticon within the digital allows it to evolve within the hyper-real into something that convinces users of its non-existence (Rheingold 1993: 289–300). Yet what is retained is a core panoptic element in which a small group can control much larger numbers. One only has to think of the power of Facebook's algorithms in the control of news flow to its users, a power not unrecognized by the platform who have sought to experiment with the control it has over the emotional state of users. Knowledge of the reach of Facebook has only belatedly reached public attention through the revelations about Cambridge Analytica who managed to harvest the Facebook data of millions of users and then sell the 'psychographic micro-targeting' of political adverts through the Facebook platform to the highest bidder. The whistle-blower, Chris Wylie, described Cambridge Analytica as being able to whisper multiple different messages simultaneously to millions of individuals with those individuals being completely unaware of what was being said to the other, a practice that fundamentally undermined the democratic principle of open debate. But, however scandalous this may seem, Cambridge Analytica were simply mirroring, perhaps in a more sophisticated form, Facebook's own business model which is based on data freely given by the user.

Transparency, democracy, and political modernity

Data strips the individual out of the decision in two ways:

First, through a mediation. To look at a data set in itself gives no knowledge, just a bunch of numbers on a spreadsheet. The problem of transparency may therefore not regard the availability of data, but our awareness of what it means. We do not understand data. We cannot read it. We cannot see it. Vast amounts of data are now collected as a by-product of electronic administration but what, if anything, a data set can tell us is not, at first glance, apparent. Data is opaque. Before its secrets are unlocked, even at the most basic level, it requires visual representation and analysis. The interpretation and visualization of data are therefore highly specialized tasks. However, a generalized lack of knowledge regarding data and statistics means that data requires mediation before it becomes meaningful. The information society is, it turns out, often illiterate when it comes to data. Data transparency can turn into agnatological production through this data illiteracy where data is wilfully used to misrepresent (Proctor and Schiebinger 2008). Not only are we reduced to the being of a data subject but our understanding of that subjectivity is, unless we understand data, always mediated and controlled. Increased data transparency, without an increased understanding of what data is and how to read it, does not lead to a legitimate form of democratic politics, just a further level of complexity. Modernity produced an explosion of knowledge to the extent that an individual can never understand it all. Modernity has doomed us all to be always in the dark relative to each other – this has only been magnified by the advent of digital technology. However, the transparent society assumes that we can see all, yet it remains opaque. Lack of ability to understand data and the lack of time available to assess all available data in each situation amounts to mass alienation. Transparency appears to imply the possession of knowledge and so, as Fenster has argued, 'transparency is a theory of communication that massively simplifies the modern state, this has the result of blinding democracy to complexity and produces a passive citizenry that is not able to work through the challenges of mass society' (Fenster 2015). Transparency is an idea born of the democratic ideal that nonetheless seems to undermine the practice of democracy.

Second, through neutralization. Transparency has come under criticism not merely for the often-mentioned sinister reasons but because it depoliticizes decision-making (Birchall 2014). This means government stripped of politics, that is, government as mere administration. Transparent decisions are assumed to be rational and not based on arbitrary (political) choice. Neutral politics, fuelled by transparent decision-making is assumed to be free from ideology and the outcome of a rational process that all can see. This is what Birchall has called 'data in lieu of politics' (Birchall 2015: 187). To situate neutralization within the era of transparency it is helpful to briefly review a lecture given by Carl Schmitt in 1929 titled 'The Age of Neutralisations and Depoliticisations'. In the lecture Schmitt conceives the preceding 400 years as four stages of neutralization of the political:

> There are four, great, simple, secular stages corresponding to the four centuries and proceeding from the theological to the metaphysical sphere, from there to the humanitarian-moral and finally to the economic sphere. (Schmitt 1993: 131)

These central spheres are successively depoliticized and rationalized so that they no longer allow room within which differences can move to the level of the political. Each central sphere represents a cultural epoch where meaning is generated from the central sphere. The epochs following the secularization of the theological began to conceive of human progress, 'The concept of progress, i.e., an improvement or completion became dominant in the 18th century, in an age of humanitarian moral belief' (Schmitt 1993: 135). In an economic age 'it is self-evident that progress is economic or technical progress' (ibid.) in which all problems are those of production and distribution. In this era moral and social questions became superfluous. Transparency has an unquestioned and pervasive meaning and it is, as Christensen and Cornelissen have argued, the mythic principle of modernity (Christensen and Cornelissen 2015). Following the enlightenment waves of modern thought spread and desacralized space by opening it up and revealing its secrets in the new public sphere. Nothing less was required for the rise of democratic institutions than the opening up of previously hidden forms of power. The logic of contemporary transparency completes this process. However, the volume of contemporary data simply multiplies complexity. This complexity makes full understanding all but impossible for a given individual even though selective use of data is routinely misused to create the impression of stability where none is present. Modernity is awash with data most of which we are always already alienated from, an existential situation which is unacknowledged.

The neutralization of the political within the transparent space follows the enlightenment legacy and the liberal principles that are encompassed within it. To quote Birchall 'Liberal democracy is deemed "fairer" simply by allowing scrutiny rather than with reference to the economic or social policies being scrutinised' (2011: 63). All questions become ones of openness while the content of a decision becomes meaningless. As long as there is transparency, the conduct of an individual becomes beside the point and so ethics is no longer important. If we can see the process, any problem is there to be seen and there can be no nefarious agenda and so this rational process removes the possibility of resistance. In this sense, transparency is the outcome of the modern conception of man as a rationally calculating being who is assumed to be maximizing self-interest. In this situation transparency is a method for individuals to police each other through a generalized panopticism and only through transparency can past infractions be punished in the market through future interactions. The assumption that we are all being watched keeps us in check. The regime of transparency therefore, the outcome of modernity, has produced a generalized distrust. We see people now for what they are, individually flawed, but without getting anyway near to a full understanding. We see the choices of an individual in public life without ever understanding the context. We can only ever be disappointed and so generalized distrust demands transparency as a panacea. Transparency holds people to account, because they need to be held to account, they are desacralized and assumed to be 'all the same' – the bar is set precipitously low.

The secret as heterotopia

In his 1906 paper 'The Sociology of Secrecy and of Secret Societies' Georg Simmel described 'secrecy as the sociological end in itself' and the secret society as 'a community for the purpose of a mutual guarantee of secrecy'

(1906: 477). What the secret society offers is 'a territory [in which] the norms of the surrounding society do not apply' (ibid.: 481). The secret society seems therefore to establish a heterotopic space that deviates from the social norm (Foucault 1986: 25). Secrecy is founded upon exclusiveness where 'whoever is not expressly included is excluded' (Simmel 1906: 490) and so the secret produces a community for those in possession of it. One is admitted to the community through the gift of the secret and the community is maintained through the secret's keeping. The secret society further establishes a heterotopic space by placing entry behind the performance of a rite (Foucault 1986: 26) through which the secret establishes a bond of trust that forms the community. Furthermore, the secret offers protection for the holders in a political, outward sense while also creating a different space of essentially un-administrable freedom within. Secrecy can therefore be constitutive of group identity (Costas and Grey 2014) and ground the political.

Whether for a political purpose or not the ability to remain hidden grants power, so much so that Simmel says 'secrecy … is one of the greatest accomplishments of humanity' (Simmel 1906: 461). Secrecy produces an enigmatic power that places knowledge out of bounds: this, for example, is the power of monarchy or the Latin Mass where the unknown element sets these figures apart. These regimes are in the open but remain hidden; they are both seen and unseen. In this sense they seem to mirror comments by Foucault on the body as a site of utopia when he says, 'Incomprehensible body, penetrable and opaque body, open and closed body, utopian body' (Foucault 2006: 231) as well as the system of 'opening and closing' that characterize the fifth principle of heterotopia (Foucault 1986).

The secret, because of its sociological value, has been employed across human cultures to impose regimes of power and to establish heterotopian spaces. Part of the liberal ideal was to break down these regimes – the liberal regime therefore might be understood as a disruption of heterotopia. The suspicion of the secret is though, perhaps, understandable. Where a secret is thought to be present those who are not party to it can presume anything of it and can project their fears and hopes on to an unknown interior. This establishes a communal problem because, as Simmel tells us, 'in the interest of association, and of social coherence, each must know certain things with reference to each other' (Simmel 1906: 455). We are back to transparency. For the primary community to function there must be reciprocal knowledge of each other; this is what democratic theory recognized.

This notion of reciprocal knowledge becomes particularly problematic within the mass society. How is it possible for there to be reciprocal knowledge of each other, how does trust develop, and how do a multiplicity of secret communities coexist within a large primary community without descending into mistrust and recrimination? This problem is particularly acute when information and knowledge develop to an extent that it is beyond a single individual's ability to know all. In this situation a division of knowledge is inevitable through which areas of expertise are gathered to specific groups.

Secrecy has a dual role that causes its problematic reception. On the one hand, it constitutes communities through a binding. On the other hand, it is a political tactic and a means to avoid the primary community and to change or control it. From the point of view of the primary community, a community bonded through secrecy that has a political role and is a threat. As monarchy shows us, the secret is a powerful affective force open to fantastical imaginings. Secret societies are prone to these same imaginings and therefore to their perceived dangerousness.

There has to be some kind of a negotiation between what is hidden and what is open. Secret spaces are needed for the creation of micro-communities and the un-administrable. But equally, democratic governance demands a certain degree of openness. As Geoffrey Bennington has shown, even Kant recognized this dilemma in the essay 'Perpetual Peace'. In Bennington's reading there is a tension between the moral politician and the political. Public right dislikes the secret but, as Bennington says, 'the transcendental injustice of political prudence is seen in the fact that it would miss its aim if it didn't keep its maxims secret' (Bennington 2011: 32). Politics is impossible without the secret but a moral politics is impossible with the secret. This is not to say that the transparent is in itself moral but that what is moral is dependent on a certain openness. This fact however that politics requires the secret must, if politics is to be moral, remain secret.

A different approach, one that doesn't revolve around a binary between secrecy and transparency, between openness and closure, is required. In light of contemporary data politics it has become more common to question transparency and to seek to inhabit the hidden space and this is what contemporary interest in secrecy seems to be doing. Julian Bratich, for example, has written about the production of secrets as a political force that produces a minor secrecy as a form of constituent power (Bratich 2007). Bratich draws on public masking, citing the tactics used by Black Bloc anarchists or Subcomandante Marcos of the Zapatistas (see Faramelli, in this book, for more on the mask and the Zapatistas). However, we cannot think of these forms of masking without mentioning others. Riot police are masked, and in London a few years ago got into the habit of removing their ID numbers while attacking protesters. The KKK and, of course, many Jihadi employ the mask. Secrecy as a political tactic is open to anyone. A minor politics can use the secret (or void as Faramelli puts it, see this book) for protection from the primary community because it seeks to disrupt and challenge the primary community and to change it, for whatever reason. This is no more than political prudence. What Bratich is really suggesting is that Left politics inhabits the secret as a strategic move in order to enact change and protect itself.

These examples engage with the secret as a purely political phenomenon and a tactic. Secrecy is necessary for a Left politics and so the Left must abandon the enlightenment goal of transparency for strategic reasons. Though this is clearly the case it doesn't get to a fundamental understanding of the secret. Birchall, on her part, cites Wikileaks as being both transparent and opaque, revealing what is hidden behind state and corporate power while, apart from the narcissism of Julian Assange, avoiding transparency themselves (Birchall 2011: 62). It is being both open and opaque which offers the most fruitful avenue from which to escape the binary.

Secrecy and non-knowledge

As part of this debate it is worthwhile returning the work of Georges Bataille. Birchall has already noted the relevance of Bataille here (Birchall 2015 2016) though in an all too cursory manner which simply cites the formation of the secret society *Acéphale*. It is difficult to comment conclusively on Acéphale, its secrecy was strongly kept so taking it as an example to follow is problematic, if a political tactic is desired. However, through Bataille's writings we can draw out an engagement with the not-known of secrecy. 'Brotherhoods, Orders, Secret Societies, Churches' was a talk given to the Collège de Sociologie in 1938. The talk was supposed to be given by Roger Caillois but he could not attend; it was instead delivered by Bataille who drew on Caillois's notes while adding his own comments. The tension between the two produced an interesting take on the cowritten text.

For Bataille and Caillois a secret society is a group that develops within another. Bataille expands by saying:

> I assume that the 'elective community' or 'secret society' is a form of secondary organisation that possesses constant characteristics and to which recourse is always possible when the primary organisation of society can no longer satisfy all the desires that arise. (Bataille and Hollier 1988: 149)

In this description secret societies rejuvenate the primary society because they are 'charged with providing a sort of active reality' (ibid.: 153) and 'collective ecstasy' (ibid.: 156). In these terms secret societies offer different spaces within which new modes of being can be enacted. The secret society provides interior purpose that is opposed to the static cohesion of the administrative world. But for Bataille it also seems to be an end in itself. In a hidden space modes of being can be enacted away from the gaze of the administrative world established by modernity. Secrecy is not here a tactic but a means of inhabiting the social which, as a by-product, produces change within the primary community. This change only occurs because the secret society exists within a space that is hidden from the primary community, but this change is not necessarily a goal.

Bataille makes a distinction between the existential secret society and one that is expressly political. The conspiratorial, political, secret society is a society formed with the express purpose of acting; the conspiratorial society has a teleology, a goal. It is a political organization that employs secrecy as a tactic. Alternatively, the existential secret society, Bataille tells us, 'designate[s] a reality existing for itself, a reality in which the pure and simple pursuit of existence, the pure and simple will to be, is what matters, regardless of any particular goal' (ibid.: 155). It is this form of secret society that Bataille was involved with through the Acéphale group as a community formed through transgression. The secret society of the headless Acéphale was the avoidance of administrative reason and was aimed at the recreation of myth. Acéphale acted as the esoteric heart of the exoteric project of the College of Sociology. The exploration of the ecstatic could take place in secret while in public the argument and the respectable, scholarly edifice could be built. In this sense Bataille seems to have understood the necessity of being both open and opaque in the construction of other spaces.

Caillois was not a part of Acéphale and he seems to have been sceptical about it. In *Preamble to the Spirit of Sects*, in which he dissects Acéphale,

Caillois calls secret societies a form of 'seduction' that begins with 'an organisation that would start by uniting a few men who are dissatisfied with the world in which they live and seek to reform it' (Caillois 2003: 208). Caillois then compares such 'reveries' as Acéphale explored with the coterminous rise of Nazism, also through the mechanism of the secret society. Caillois seems to have been concerned with the political ambiguity of the secret as useful to Left as to the Right. Bataille was also aware of these tensions within the secret society and sought to explore them, which perhaps explains the important distinction between the existential and political secret.

Caillois's notes on secret societies do however contain a third route for the elective community though his remarks on brotherhoods. A brotherhood is both secret and public. Its presence is not hidden but its membership and activities can be. Brotherhoods are elective communities that are based on a hidden element though they remain public (Bataille and Hollier 1988: 151). They are therefore a locus of prestige that openly acknowledges the power of anonymity that embodies the diffuse power of a division of knowledge.

For Caillois, the medieval guild is paradigmatic of the brotherhood. The elective community is formed here through the possession of technical knowledge or language. The social realm is replete with such brotherhoods, though they are often viewed with suspicion. John Lanchester, for example, has detailed the use of opaque language of finance (also see Marazzi 2011: 123–35; Lanchester 2015). Language here is taken as code that potentially hides meaning while remaining, to a certain extent, transparent – but what is being described by seemingly obtuse language (often wilfully so) are highly technical processes. Medicine is another one. We should not forget that this is also a common complaint about much academic language and is also a key to the misuse of statistics. But in these modern brotherhoods the languages are learnable and give access, though they may be inherently complex and difficult. Caillois's understanding of brotherhoods is meritocratic, and secrets here are not conspiratorial. By understanding these specialisms as meritocratic brotherhoods we can see how they can cohere within a primary community both openly and opaquely. Another example of a brotherhood, as this form of elective community, can be found on dark web social networking. Dark web social networking retains anonymity both through what users reveal of themselves – giving out personal information is taboo – and in the means of access. The means of access makes dark web social networking self-elective and in this it is revealing. One has to make an effort to access the dark web – it is not a simple matter of using a search engine. Access requires effort and technical skill which acts as what users self-describe as an 'admission test' that produces a 'techno elite' (Gehl 2014). What is significant here is that a dark web techno elite avoids contemporary transparency and surveillance while retaining the human connectivity of the web that demands active subjects in membership. This picture of dark web social networking is opposite to the common understanding of the dark web which is perceived as the realm of terrorists, drug dealers, thieves, and paedophiles. Outside of the elective community of the dark web there is the blank space of the secret which is open to a multitude of interpretations.

The understanding of IT, modern finance, medicine, statistics, or critical theory, as well as car maintenance or plumbing are technical tasks that require time to learn; they embody expertise. A mass society cannot function without such a division of knowledge and it is impossible for all to know everything. A certain amount of opacity is inevitable because of complexity but the ideology of transparency pretends that this is not the case. We could potentially learn these arts but it is an unfeasible option. What is required is a certain amount of trust between the micro-communities and the primary community. Learning the language of computer code, finance, or whatever, gives us access, if we want it, but we cannot all gain access to all brotherhoods. The presence of the secret tells us that we cannot know everything. Learning to inhabit the secret should not therefore be about using the secret for political goals but about accepting a fundamental non-knowledge as being ontologically and sociologically constitutive. The secret is therefore encountered not as a problem but as a constitutive aspect of social reality which mirrors the ontological condition of non-knowledge.

If Acéphale was, in its own way, a political project, Bataille's subsequent work moved beyond even this semblance of outward facing activity. Bataille began to describe what he called 'non-knowledge'. He described non-knowledge not from an epistemological perspective as a mere lack of knowledge but as a discomforting experience. As a lack, non-knowledge becomes the absence of knowledge and a problem to be overcome, one that, following the logic of rationalism and transparency, *can* be overcome. For Bataille, however, non-knowledge is an experience that is deeply unsettling, primarily because it *cannot* be overcome:

> To specify what I mean by non-knowledge: that which results from every proposition when we are looking to go to the fundamental depths of its content, and which makes us uneasy. (Bataille 2001: 112)
>
> I myself am in a world I recognise as profoundly inaccessible to me … I remain in a kind of despair. … This is the position of someone who doesn't know what is in the locked trunk, the trunk there is no possibility of opening. (Bataille 2001: 113)

Bataille pinpoints the radical discomfort of not knowing which the secret society is an element of. Post-Hobbesian political thought posits that the role of the sovereign is to enable the reduction of pain and the maximization of pleasure. In these terms the horrendous condition of non-knowledge is there to be negated and this drives the quest for absolute knowledge. In a 1955 essay 'Beyond Seriousness' Bataille puts it in this Hegelian context:

> Hegel hasn't grasped a singular inhumanity in our completion, the absence of all seriousness in the man finally liberated from the *consequences* of

> servile tasks, no longer having to take these tasks seriously, no longer having to take *anything* seriously. (Bataille 2001: 214; emphasis in original)

Bataille's critique of Hegel is that he produces a system which he intentionally closed and made servile:

> Hegel, I imagine, touched upon the extreme limit. He was still young and believed himself to be going mad. I even imagine that he worked out the system in order to escape (each type of conquest is, no doubt, the deed of a man fleeing a threat). To conclude, Hegel attains satisfaction, turns his back on the extreme limit. Supplication is dead within him. (Bataille 1988: 43)

Bataille's claim is that Hegel encountered the limit experience of non-knowledge and turned away from it, 'shrinking away from ecstasy … he had to take refuge somewhere' (Bataille 1988: 108). The system of knowledge attempts to turn the experience of non-knowledge, which Bataille insists is inherently uncomfortable, into a lack through which it can be negated. Hegel's system is an attempt to negate this discomfort. In this context we can understand transparency, as Sam Webber has noted, as 'a response to the anxiety of not being able to occupy a secure, stable, identifiable position in the world' (Phillips 2012: 162). For Bataille, boxes can be opened but within them, there are only more boxes. The era of transparency can only continue to endlessly expose more, create more data, open up more sites for viewing, forever becoming more transparent. This pursuit of knowledge, the grounding of modernity, is an ultimately futile task:

> I think that knowledge enslaves us, that at the base of all knowledge there is a servility, the acceptation of a way of life wherein each moment has meaning only relation to another or others that follow. (Bataille 2001: 129)

This is accepted because the problem of non-knowledge is accepted. Because non-knowledge is accepted as an uncomfortable lack that should be negated. Instead, for Bataille, non-knowledge should be experientially and bodily inhabited. Laughter, tears, and what Bataille describes as inner experience form part of what he calls an atheology of affect that amounts to a refusal of project and modernity's fetish of progress. The secret society of Acéphale cannot be understood apart from this experience of non-knowledge and its antimodernist trajectory. Bataille's thought contains a rediscovery of the Christian tradition of mysticism and the lived experience of the sacred in light of modernity's calamity in the 1930s and 1940s (Connor 2000). Non-knowledge moves from a historicist problem to one that is experienced in the body, that is, it spatializes it. Non-knowledge is something that should be experienced and not negated dialectically.

This is not to argue that the search for knowledge and a democratic desire for transparency are entirely wrong-headed, nor to argue against technological development. However, learning to inhabit non-knowledge as an experience disrupts the illusion of progress and questions the foundational premise of modernity through an ethical response. The question asks, what are knowledge and transparency for? The misprision of modernity is to take them as ends, this is how transparency neutralizes the political. However, an increase in knowledge and an increase in transparency are not the same as an increase in ethical being, they are technical responses to an existential question. Rationalism, as the guiding principle of modernity, leads inextricably towards Hegel's administrative state, the present outcome of which is a neoliberalism that is justified as an appeal to the modern as rational. Laughter, tears, and eroticism as limit experiences and embodiments of non-knowledge question the triumph of this rationalism and present an alternative mode of being that accepts the presence of the not-known. The secret community is one mode of doing this.

Bataille's understanding of love will provide a useful example. Love, for Bataille, is only possible as a manifestation of excess. It flourishes within an abundance of resources and, importantly, 'clandestinity is not at all necessary to individual love, but it often increases the intensity of feelings'

(Bataille 1999: 157). Love transcends the useful in society, the goal of love is not procreation, nor is it about social calculation; the lovers are an end in themselves. Love transgresses the rational modern state because it is not based upon rational self-preservation. Despite capital's attempts to commodify it, love is entirely self-sufficient, relying only upon the excess of the lovers. This is particularly so when we think of it as a form of secret. The lifeworld of the lovers is a secret community born of intimacy. Indeed, when we look at attempts to commodify love they often revolve around its becoming public, the lavish wedding, for example, or the ostentatious valentines gift. For Bataille, in marriage 'the union [of the lovers] is stabilized, at least in appearance. The sexual play of the lovers has reproduction and growth of a family as its effect, if not as its purpose' (Bataille 1999: 163). The open occurrence of love in marriage acts as a veil of purpose upon the private world of excess. The lovers point to the actuality of the secret, on the one hand, the private world of excess gives the community its meaning but, via marriage, it limits itself in the eyes of the wider community. The lovers therefore seem to offer a model of how to inhabit the secret within the democratic space by being both open and opaque.

The lovers embody a further hallmark of the secret in trust. The experience of love like the rest of what Bataille understood as inner experience is internalized and unspeakable – much of Bataille's project was an attempt to understand this unspeakability of inner experience – however, the lovers are plural and rely upon mutuality. Each must believe that the inner world of the other bears a resemblance to their own experience of love. Necessarily, this cannot be known and so the lovers must trust each other. As Simmel pointed out 'absolute unity' (Simmel 1906: 460) is impossible, that is, transparency for the lovers is not an option so 'the secret of the one party is to a certain extent recognised by the other … and respected' (ibid.: 464). Trust, for Simmel, is the hallmark of the secret. The secret presents a reciprocal confidence of those party to it – especially so in the secret society where the members must have faith and a moral accountability to each other. Trust is a compensation for, and acceptance of, incomplete information which seems to be based upon the acceptance of the bodily experience of non-knowledge through its discomfort. Trusting is difficult, and sometimes painful and discomforting. The person who trusts opens themselves to wounding and laceration.

It is through this notion of trust that we can navigate through the apparent demand of democracy that we have transparency. Democracy, as living together equally, demands trust as much as it demands openness. It is as impossible for the lovers to achieve absolute unity as it is for the democratic community. It is not possible to possess the necessary information regarding any democratic choice which is why sovereignty is entrusted to representatives. The democratic system can only function if the community manifests partially through trust. Accepting a lack of knowledge, just as accepting the secrets of a lover, begins to constitute that community. But it is exactly this element of trust that breaks down within neoliberalism. Based upon the reductive logic of homo economicus it is understood that each individual is concerned only with their own self-interest, a logic that is captured in economic modelling. The rationality of neoliberalism, which derives from the Hobbesian premise that overturned classical thought, understands the human as rationally pursuing the avoidance of their own pain and maximizing of their own pleasure. The only trust that can exist is that each individual is maximizing their own self-interest, we can trust in their being no trust.

In this situation, the logic of transparency becomes obvious. Transparency is necessitated by the understanding of the human within modernity which presupposes a breakdown of trust. Inhabiting secrecy may therefore be, counter-intuitively, a method through which trust can be reacquired. Accepting what is beyond knowledge and embracing the discomforting feeling of the unknown forces one to encounter and inhabit non-knowledge rather than positing a lack which must be overcome through technical means. Encountering non-knowledge in this way can force democratic politics to accept that it cannot know and monitor all and that a desire for knowledge is untenable on these terms. Furthermore, it forces one to encounter that a mass society only functions with a division of knowledge that necessarily produces secrets. The division of knowledge, like non-knowledge, should be encountered as an experience and not, as the market solution suggests, a lack to be overcome.

Conclusion

At the heart of being there is a fundamental opacity. The project of modernity questions this by turning the dark heart into a simple lack to be overcome. The unknown is therefore turned into a threat and not a mystery. The desire to see all turns into a demand for everything to open up and the false assumption that all can be known. Though knowledge and accountability have been, and still are, beneficial, the project itself is Sisyphean. It is by inhabiting secrecy that we can live in the world with non-knowledge. Not, as may be the first thought, part of a political strategy to avoid political disruption but as a model for existing within incomplete knowledge. In this sense the secret space stands apart as it relies upon trust for its maintenance as well as for its acceptance by the outside.

Inhabiting secrecy through the experience of non-knowledge seems to present a critique of contemporary rationalization because it questions the teleological assumptions of modernity. Non-knowledge is not posited as a lack to be progressively overcome but a condition of being, an ever-present with no future resolution. The secret is not here inhabited as political tactic but as an end in itself and as an experience of fundamental non-knowledge.

Inhabiting the experience of non-knowledge faces one towards the impenetrability of the world and forces one to make an accommodation with the limited nature of existence. This allows one to be both open and opaque. This can be achieved communally through trust where absolute unity is accepted as an impossibility and where a certain level of opacity is accepted and experienced. The inhabited secret negates the logic of transparency which assumes the lack of knowledge can be broken down to allow a form of policing in lieu of trust. By inhabiting the secret and by living the experience of non-knowledge we can begin to comprehend the complex division of knowledge that mass society requires but without the teleological assumption that the lack of knowledge can be overcome.

The secret refutes the very premise of liberal governmentality, of transparency and sameness, but only if it is inhabited rather than employed as a strategy to enact a particular political programme. This leads to a twofold need.

First, a re-appreciation of the right to opacity and privacy, others have already argued for this (Bratich 2007: 53). At the micro level what has become known as digital disengagement may be an important strategy and has been described by Kuntsman and Miyake in a developing research programme as a 'proactive form of citizenship' (Kunstman and Miyake 2015). Digital disengagement involves removing oneself from the transparent space in favour of non-digital spaces and through social relations that are non-mediated. In their preliminary study Kuntsman and Miyake produce an ethnography of digital disengagement that describes an interaction between the individual and the transparent space. They point to the negotiation that these individuals undertake involving their concerns about the digital space, primarily privacy and social relations, and the requirement of contemporary cultural economy to embody a digital subjectivity. This negotiation forces individuals to leave but then return to the digital space in order to function within the wider socio-economic realm. What seems to develop is a critical digital subjectivity that is both open and closed, that limits transparency by curating a space that is hidden from the panoptic gaze of the digital by limiting digital exposure. This may mean, for example in the case of turning off Google maps, making a journey inefficient but it also acts as a refusal of economic rationality which is the core of the devil's bargain with digital transparency. This is not difficult to do but it might be difficult to maintain, removing oneself from social media applications means cutting oneself off from contacts, colleagues, and comrades but with the knowledge that there exists a space in which one is ostensibly hidden. Critical digital subjectivity might therefore be a way of curating opaque spaces beyond the panoptic gaze that are unknowable except through genuine social relations, open but opaque. Inhabiting opacity in this way would create a space outside logic of economic reason and establish heterotopias of genuine intersubjectivity.

Second, a re-appreciation that communally held secrets amount to Caillois's understanding of brotherhoods and as centres of prestige and expert knowledge. The acceptance of non-knowledge comes with an acceptance of the diffuse elements of power within the division of knowledge and this in turn leads to the repudiation of what Fenster noted as the simplification of reality which transparency brings. This does not mean closed communities but ones that are open to membership on meritocratic lines. We must accept exclusion from communities that we do not have time to understand and thus acknowledge the political rather than neutralize it. Equally, exclusive communities have to grasp that their position within the primary community and their opacity necessarily causes suspicion. To hold a secret openly is to ask for the trust of the community within which it exists.

The secret creates an ethical demand that transparency pretends is not there. If we admit to the presence of the secret and the status of non-knowledge we disclose the fact that the social whole can only be maintained through trust. The need for trust places a demand upon the community holding the secret. If the knowledge of the community grants it power, for example, the knowledge of medicine, of statistics or of financial services, there is a demand that that knowledge be employed wisely. Secrecy admits this. Transparency, on the other hand, presumes that all knowledge is used against the interests of the wider community and can only be policed through transparency. The latter is an impossible task.

Note

1 For an accessible summary of the business model of Facebook, see (Lanchester 2017).

References

Bataille, G. (1988), *Inner Experience*, Albany: State University of New York Press.

Bataille, G. (1999), *The Accursed Share: An Essay on General Economy. Vol. II. The History of Eroticism. Vol. III. Sovereignty*, New York, NY: Zone Books.

Bataille, G. (2001), *The Unfinished System of Nonknowledge*, Minneapolis MN: University of Minnesota Press.

Bataille, G. and Hollier, D. (1988), *The College of Sociology (1937-39)*, Minneapolis: University of Minnesota Press.

Bennington, G. (2011), 'Kant's Open Secret', *Theory, Culture & Society*, 28(7–8): 26–40. doi:10.1177/0263276411423036.

Birchall, C. (2011), 'Transparency, Interrupted: Secrets of the Left', *Theory, Culture & Society*, 28: 60–84. doi:10.1177/0263276411423040.

Birchall, C. (2014), 'Radical Transparency?' *Cultural Studies ↔ Critical Methodologies*, 14(1): 77–88. doi:10.1177/1532708613517442.

Birchall, C. (2015), '"Data.gov-in-a-box": Delimiting Transparency', *European Journal of Social Theory*, 18(2): 185–202. doi:10.1177/1368431014555259.

Birchall, C. (2016), 'Managing Secrecy', *International Journal of Communication*, 10(1): 152–63.

Bratich, J. (2007), 'Popular Secrecy and Occultural Studies1', *Cultural Studies*, 21(1): 42–58. doi:10.1080/09502380601046956.

Caillois, R. (2003), *The Edge of Surrealism: A Roger Caillois Reader*, Durham: Duke University Press.

Christensen, L. and Cornelissen, J. (2015). 'Organizational transparency as myth and metaphor' European Journal of Social Theory, 18(2): 132-149.

Coll, S. (2012), 'The Social Dynamics of Secrecy: Rethinking Information and Privacy Through Georg Simmel', *International Review of Information Ethics*, 17: 15–20.

Connor, P. T. (2000), *Georges Bataille and the Mysticism of Sin*, Baltimore: Johns Hopkins University Press.

Costas, J. and Grey, C. (2014), 'Bringing Secrecy into the Open: Towards a Theorization of the Social Processes of Organizational Secrecy', *Organization Studies*, 35(10): 1423–47. doi:10.1177/0170840613515470.

Fenster, M. (2015), 'Transparency in Search of a Theory', *European Journal of Social Theory*, 18(2): 150–67. doi:10.1177/1368431014555257.

Foucault, M. (1986), 'Of Other Spaces', *Diacritics*, Baltimore: The Johns Hopkins University Press, 16(1): 22–7. doi:10.2307/464648.

Foucault, M. (2006), 'Utopian Body', in C. A. Jones (ed.), *Sensorium. Embodied Experience, Technology, and Contemporary Art*, 229–33, Cambridge, MA: MIT Press.

Gehl, R. W. (2014), 'Power/Freedom on the Dark Web: A Digital Ethnography of the Dark Web Social Network', *New Media & Society*, online fir, 1–17. doi:10.1177/1461444814554900.

Kunstman, A. and Miyake, E. (2015), *Paradoxes of Digital Dis/Engagement: A Pilot Study (Concept Exploration)*. Available at: http://www.communitiesandculture.org/projects/connecting-communities-paradoxes-of-digital-disengagement/ (accessed January 2018).

Lanchester, J. (2015), *How to Speak Money: What the Money People Say and What it Really Means*, New York: W. W. Norton.

Lanchester, J. (2017), 'You Are the Product', *The London review of books*, NYREV Inc, 39(16): 3–10. Available at: https://www.lrb.co.uk/v39/n16/john-lanchester/you-are-the-product (accessed: 8 August 2017).

Lannier, J. (2013), *Who Owns the Future?* New York: Simon & Schuster.

Marazzi, C. (2011), *The Violence of Financial Capitalism, New* edn. Cambridge, MA: semiotext(e).

Phillips, J. (2012), 'Secrecy and Transparency: An Interview with Samuel Webber'*Theory, Culture & Society*, 28(7–8): 158–172.

Proctor, R. and Schiebinger, L. L. (eds) (2008), *Agnotology: The Making and Unmaking of Ignorance*, Stanford: Stanford University Press.

Rheingold, H. (1993), *The Virtual Community: Homesteading on the Electronic Frontier*, Reading, MA: Addison-Wesley Publishing Company.

Schmitt, C. (1993), 'The Age of Neutralizations and Depoliticizations(1929)', *Telos*. Telos Press, (96): 130–42. doi:10.3817/0693096130.

Simmel, G. (1906), 'The Sociology of Secrecy and of Secret Societies', *American Journal of Sociology*, 11(4): 441–98. doi:10.1086/211418.

6

Mirrors and masks: The political space of Zapatismo

Anthony Faramelli

Introduction

Throughout his oeuvre, Michel Foucault radically worked to rethink spatiality in politics and political philosophy. However, rather than focusing on large-scale geopolitics, Foucault was, famously, primarily concerned with the study of 'other spaces', spaces where reality was inverted or distorted. As a means of thinking through these other spaces, Foucault coined the term 'heterotopias' to signify other spaces that effectively enact counter zones in which real sites found within a culture are simultaneously represented, contested, and inverted. In this way Foucault remarkably foresaw both the spatial organization of neoliberalism[1] and the spatiality of resistance to neoliberal empire. In 'Of Other Spaces: Utopias and Heterotopias' Foucault wrote that the concern with space in the West has a history of conflict and contestation dating back to Galileo's 'scandalous' rediscovery of infinite and infinitely open space, effectively contesting the church's domination of space as enclosed spaces that can be extended to localized points within an infinite space (1986: 22–3). For Foucault it is the accumulation of the history of spatial contestation that gave rise to the polemic ideological conflicts that sought to control and enclose space in the late twentieth century (ibid.: 22).

While 'Of Other Spaces' touches on multiple heterotopic inversions such as disciplinary spaces of deviation, perhaps the most intriguing other space Foucault engages with is the space of mirrors. While this has not generated as much interest or critical engagement as his writings on prisons or asylums, the mirror invites us to reflect on a space of mediation that can be utilized in the service of micro-political struggle. This chapter will accept Foucault's invitation and expand on the latent critical mediating force present within mirrors' inverted space. It will then examine the way that this space has been utilized by the Zapatista resistants as a vehicle of critical response to crisis. Finally, following the work of Friedrich Kittler, this chapter will argue that this technology of mediation is in fact central to the Zapatista subjectivity. Through its examination, this chapter will argue that the mirror, as a heterotopic technology of mediation, has become the key instrument in forming an actionable critical response to the omnipresent crisis of neoliberal economics.

Mirrors and the construction of reality

For Foucault the mirror opens up a virtual space where people, paradoxically, see themselves in a space where they are absent. In this sense the mirror is heterotopic insofar as it is real, but the space does not exist in reality. It places the person looking into the mirror into the virtual realm of possibilities, allowing them to see how they could be somewhere else (1986: 25). It is worth quoting Foucault at length:

> The mirror is, after all, a utopia, since it is a placeless place. In the mirror, I see myself there where I am not, in an unreal, virtual space that opens up behind the surface; I am over there, there where I am not, a sort of

> shadow that gives my own visibility to myself, that enables me to see myself there where I am absent: such is the utopia of the mirror. But it is also a heterotopia in so far as the mirror does exist in reality, where it exerts a sort of counteraction on the position that I occupy. From the standpoint of the mirror I discover my absence from the place where I am since I see myself over there. Starting from this gaze that is, as it were, directed toward me, from the ground of this virtual space that is on the other side of the glass, I come back toward myself; I begin again to direct my eyes toward myself and to reconstitute myself there where I am. The mirror functions as a heterotopia in this respect: it makes this place that I occupy at the moment when I look at myself in the glass at once absolutely real, connected with all the space that surrounds it, and absolutely unreal, since in order to be perceived it has to pass through this virtual point which is over there. (ibid.)

Foucault was not simply being poetic or allegorical when he noted the importance of mirrors joining utopias to reality. By reflecting an inverted image of the world within the mirror, the subject is able to observe and critically assess the world. The mirror reflects back the external world as it appears without *perceived* ideological filters, although the image is only ever partial, cut off by the borders of the mirror. As such the image in the mirror is true, but not the whole truth. In this way we can say that the heterotopic power of the mirror is its ability to mediate how the world is seen. This allows us to expand Foucault's concept of the heterotopic mirror to include photography, television, cinema, and even newspapers and other forms of media which create internal (to the subject) affective spaces by connecting and reflecting real-world external sites. For this reason mirrors are heterotopic sites of *mediation* where the world that is reflected back is inverted by way of framing. Heterotopias of mediation differ from utopias because they do not reflect a perfected image of society, only an *Other* image of society.

Heterotopias of mediation are in essence sites of contest where a symbolic or ideological counter-reality is created through the process of framing. Perversely, while Foucault envisioned heterotopias as spaces of possibility, the symbolic framing of these sites forecloses potentiality. At the discursive level, they are formed through *statements* that purport to have a singular operative reality that freezes all movement and critique (Deleuze 2006: 3). Deleuze, in his reading of Foucault, outlined the process through which these heterotopic counter-sites become the hegemonic site of 'reality' or reality as such, through the repetition of statements. Statements are rare and their repetition by different agents in time builds up and forms the essential structure of reality as an effect of their rarity and regularity and in this way that they are networked to each other to form a reality (ibid.: 3–4). This reality has no connection to an origin or to a return to origins, but it 'preserves itself within its own space and continues to exist while this space endures or is reconstituted' (ibid.: 4). In other words, reality as such is partial, static, and immoveable – *essential*. In this way, we can further complicate Foucault by saying that a heterotopia of mediation is not *a* mirror, but several mirrors, a hall of mirrors, where power is formed by the repetition of the same set of statements.

Importantly, this focus on discourse is not deconstructive insofar as deconstruction asserts that nothing is outside the text. For Foucault there is a grounded phenomenological experience of the world that the mirrors invert. Here Foucault is closer to Lacan and Kittler than he is to Derrida. He asserts that there is a phenomenological experience of the world that is over-coded and restructured through discourse. This entails a necessary, and potentially psychotic, schism between a person's internal space, their individual experience of the world, and external spaces reality as such.

Building on Mach and Freud's uncanny experiences of not recognizing their reflections in mirrors, Friedrich Kittler takes this point further by arguing that mirrors open up an Other world by mediating reality back to us (1997: 91–3). Kittler later goes on to write that

> mirrors are a medium not for recording, but for transmitting nature. They perform only the function of … perception, and at least in Lacan's inhuman model, this is already sufficient to cut through 'the metaphysical … problem of consciousness' with a 'Gordian' sword-stroke. More precisely, in a 'materialist definition' of consciousness any 'surface' suffices where the refraction index biuniquely transfers individual points in the real to corresponding though virtual points in the image. So-called Man, distinguished by his so-called consciousness, is unnecessary for this process because nature's mirrors can accommodate these types of representation just as well as the visual center in the occipital lobe of the brain. (ibid.: 131)

There is a certain type of violence to this form of mediation. A psychic violence (the experience of the uncanny) that flows from the mind's inability to grasp the machine. Indeed, 'the psychic apparatus', Kittler argues, 'obstructs any understanding of the technical' (ibid.: 95). This means that the metaphysical 'problem of consciousness' that Kittler references is the problem of the Lacanian imaginary. The imaginary, constituted through what Lacan referred to as the 'mirror stage', is the formation of a divided subject when an infant cannot rationalize its narcissistic image of itself (its ego-ideal) with the reality of its form. In his reading of Lacanian psychoanalysis, Kittler locates this moment – the moment of (mis)recognition when design is no longer of consequence and consciousness is tied to the contingent presence of eyes (to analogue media) – as opening a 'gap that makes room for war, tragedy and cybernetics' (1997: 143). It is within this gap that the repetition of statements, what Kittler refers to as the positive feedback of information within a knotted circuit where the repetition of media statements begins to function independently of subjectivity, begins to take on an existence as reality (ibid.: 143–4). What this means is that reality is a machinic function of media. This is what Baudrillard termed a *hyper-reality*, a reality that is constructed without a referential being or a substance, a real without origin or reality that is created by the process of mediation (2006: 1).

To take this point further still, the affective experience of the mirror's hyper-reality is exactly the uncanny experience which Kittler noted. The mirror occupies a curious and paradoxical space that is both inside and outside the subject, what Soja termed a *Thirdspace* (See Soja 1996). In other words, the mirror as a heterotopia of mediation occupies three distinct spaces, defined by the subject's position or proximity. The first space is internal. This is the space where mediation is at the heart of creating reality and subjectivity. Within the first space the mirror is imperceivable. Its function is generative of power through the repetition of statements. The second space is external and utopic. It is a space of desire where the subject can see itself where it is not, but *could be*. The second space of a mirror is analogous to what Deleuze and Guattari term the plane of consistency, a virtual space of possibilities formed through communal desire (2002: 157, 169). The mirror's third space (Thirdspace) is critical. This space cuts through hyper-reality and utopic spaces of desire, allowing a critique of both. These spaces are very much living. This is not, as Julian Reid reminds us in his contribution to this collection, simply the spaces in which we live, but a *space-image* where imagination lives, which draws us outside ourselves and allows us to travel to exterior spaces (Reid 2018: 45).

Zapatismo: Mirrors and the construction of an other Mexico

In her contribution to this book, Claire Colebrook looks at how 'Mexico' has been dialectically constructed in American discourse as a 'dark' and chaotic space: a 'land of crime, drugs, weapons, lawlessness, violence and destitution – the "Mexico" that would need to be walled off to make America great again' (Colebrook 2018: 115). Colebrook is specifically focused on Donald Trump and US Republican discourse during the 2016 presidential campaign which sought to 'make America great again' by constructing a quasi-theological Manichean opposition between America and Mexico where Mexico as a space of dark lawlessness is the necessary condition that allows the possibility for 'America' to be made 'great' again (ibid.).

Strikingly, the same Manichaean opposition has also been employed by Mexican politicians to justify unpopular economic and social reforms. Within Mexico this dialectic can be traced back to the nineteenth century when the dictator Porfirio Díaz came to power. When Díaz came to power in 1877 Mexico was almost bankrupt and social unrest prevailed, prompting Díaz to remark, 'Poor Mexico, so far from God, and so close to the United States' (Lucero 2002: 263). The Manichean opposition was obvious from the start. Mexico was a dark, lawless land while the United States was a land of light, wealth, and opportunity. However, with proper participation in a deregulated global economic order, Mexico could be the United States. The United States became an aspirational object for Mexico. To this end, Díaz embarked on a series of reforms that radically opened the country to free trade and foreign investment, particularly investment from the United States (ibid.). The reforms restored growth, but the wealth either flowed back to foreign investors or was centralized in a small handful of powerful families. Rural, poor, and indigenous communities conversely suffered further impoverishment. Their suffering lead to further unrest which finally forced Díaz out in 1911 when the Mexican Revolution began with Villa in the north and Zapata in the south.

The Mexican Revolution prevented Mexico from becoming 'America', locking the dialectic in for generations to come. However, the American aspiration lived on in the ruling elite, and was fully revived in the 1990s when President Salinas negotiated the North American Free Trade Agreement (NAFTA). The language Salinas, a Harvard-educated economist, used was a direct reflection of Díaz. Indeed, Salinas even quoted Díaz (incorrectly attributed to Juarez) when speaking to *Fortune Magazine* in 1992.[2] In the

Fortune article, Salinas stated that neoliberal economic reform needed to be locked in at the *institutional level* so it would be impossible for future presidents to reverse course (Salinas 1992). To accelerate this, the Salinas government ended the *ejido* system of communal farming that was guaranteed by the post-Revolution constitution. This was literally intended to make Mexico, the dark land of rural farmers, become (North) America, the land of light, law, and economic progress.

This is why the Zapatista Army of National Liberation (*Ejército Zapatista de Liberación Nacional* or EZLN) chose 1 January 1994 – the same day NAFTA came into effect – to launch their uprising. The Zapatistas of the EZLN emerged as a clandestine military organization composed primarily (but not exclusively) of indigenous and rural people. However, they quickly set themselves apart from the Marxist guerrilla movements that had become common in Latin America throughout the 1970s and 1980s by their refusal to take power, instead claiming that their aim was to create a *space* for democracy.[3] Subcomandante Marcos, their spokesperson during the early days of their uprising, poetically wrote that they sought to do this by opening up a 'crack' in reality from which they can critically interrogate the world in order to open up 'another world' of possibility (See Ponce de León 2001: 209–12; 18). Following this, the Zapatista resistance has been mainly concerned with the production of space, specifically the space of Mexico, a Mexico outside of the neoliberal horizon and beyond its geopolitical borders.

Zapatismo (the Zapatista political philosophy) is not interested in Manichean oppositions that place the construction of *a* Mexico against *a* United States; rather zapatismo is concerned with multiple constructions of Mexico. The Zapatista writings, especially those credited to Subcomandante Marcos, productively look to the role mirrors play when thinking through the spatiality of Mexico. This is most explicit in an essay dated June 1995, 'Power as a Mirror and an Image' (Marcos 2004: 153–60). While this essay is far from the only one to employ mirrors in its analysis, it is by far the most crucial in demonstrating the heterotopic aspects of the mirror and the role spaces of mediation play in politics. The essay is broken down into four chapters where each chapter represents a mirror. As the narration passes through each mirror, we can see that all three aspects of the mirror's heterotopic space are operative; the way mediation serves power through the internal and invisible repetition of statements, the way in which it can be situated outside the subject to offer a utopic perspective, and the mirror's Thirdspace of critique. The essay sits within Thirdspace in order to form its critique of Mexico and ends by briefly looking at how the mirror can move into the centre of a resistant subjectivity with a living utopic counter-image.

In writing that is characteristically poetic and performative, the first section, or 'The First Mirror', of this essay (ibid.: 153–6) is a critical analysis of how 'Mexico' moved from a 'mirror', a heterotopic space of mediation, to an 'image', a simulacrum of neoliberal governmentality. This chapter is divided into two sections, each section dubbed a deceit; however, a close reading shows that this chapter engages with four distinct spatial levels of analysis. The first space is a heterotopia of mediation:

> In the [neo-liberal] Power the mirror reflects a double image: what is said and what is done. The mirror hides nothing. The resources are gone, it is not the same as before. Its surface is mildewed and stained. It can no longer 'reverse' reality. On the contrary, it shows the contradiction. But in making this evident, it controls it and puts it at its service. Now it simply attempts to make the contradictory image seem 'natural', as 'evidence', and 'unquestionable'. (Marcos 2004: 153)

Using the events of 1994 and the first half of 1995 as its evidence, this chapter points out that neoliberal governmentality is fundamentally contradictory (ibid.: 153–4); however, the neoliberal state frames the contradiction in such a way as to solidify its power. It does this by positioning economic, political, and social instability as existing outside the frame of governmentality, rather than as integral aspects of neoliberalism, reflecting the statements that it is 'I or fascism', 'I or anarchy', and 'I or chaos' (ibid.: 154). In doing so, 'Mexico' is no longer a heterotopic space of mediation, but becomes a 'real' space that is more real than reality. It becomes a different space; it becomes an image, a simulacrum, where movement is frozen and critique curtailed. Returning to Colebrook's analysis, what is implicit in this discourse is the adoption of the US construction of Mexico as inherently lawless and chaotic. For this operation of Power, in the beginning that world was Mexico, but by adopting the neoliberal framework Mexico can effectively become 'America', a space of order that is guaranteed by neoliberal governmentality.

This is a space that is fundamentally tautological (Marcos 2004: 155). However, zapatismo argues that mediation alone is not sufficient to freeze the image. The movement to an image needs codification of the statements of Power; it needs legality to structure its territory (Marcos 2004: 154–5). The most dramatic event of legality used to structure the space of Mexico happened in 1994 when NAFTA came into effect. Writing on the structuring effect of legality, Subcomandante Marcos noted that legality is the

> Power: the tautological mirror. In its image, in the reflection it obtains itself, the Power says:
>
> '*I exist because I am necessary,*
>
> *I am necessary because I exist,*
>
> *therefore:*
>
> *I exist and I am necessary.*'
>
> As the image it receives of itself is enough to satisfy itself, the Power believes it is enough, and once again the mirror in front of the mirror, eternal. (2004: 155, emphasis in original)

The problem is that despite having legality on its side this image is nevertheless illegitimate (ibid.). As Sam Weber points out, there is a schism between the social contract, the pact between a people and a state that gives the state its legitimacy, and the law (1992: 245–6). Drawing on the work of Rousseau, Weber takes us to a spatial analysis of the formation of a state by noting that its 'body politic' is formed by the social contract and given life by its ability to 'move', change, and reconstitute itself through the process of legislation (ibid.):

> 'Preservation' does not follow upon the birth of the self: it constitutes it. Society must *reproduce* itself, must repeat itself, in order to *be*. But its reproduction/repetition does not add up to an integrated, self-identical state. The books do not balance. This imbalance is recounted in a narrative or origins which appears to establish a causal necessity. (ibid.: 246, emphasis in original)

Weber states that it is this imbalance that allows the legislator's legitimacy to be questioned, leaving the social contract to be able to do nothing more than 'engender its own, invented mirror-image ad infinitum', rendering the law nothing more than a 'simulacrum' (ibid.: 249).

However, if, as Weber argues, it is the social contract that gives birth to a people and legislation that gives life to a people, then we must ask what people have been given life and which people are left in a tautological 'dead' state? The adoption of NAFTA was intended to bring movement, life, to a certain class of people, leaving many others trapped in a static system. As such, the answer we are left with is that the simulacrum is not total. What is created is a double articulation of Mexico, two Mexicos, one alive and active in the global neoliberal system and another dead space of stasis.

The signing of NAFTA was intended as the move to 'make Mexico America', to fully integrate Mexico into the American neoliberal system. However, as Colebrook points out, this is an image that is solely internal. Externally, in America, Mexico is still constructed as a dark and chaotic territory (Colebrook 2018). This image of Mexico is also reflected to the

Mexican people via foreign press (Marcos 2004: 154). However, it is *crisis*, specifically the political and economic crises of 1994 and 1995, which opens up a heterotopic space for a critical analysis of neoliberal governmentality (ibid.). It is the lived, phenomenological reality of crisis that opens the cracks in the simulacra of Mexico for those living in the tautological, dead, Mexico.

Marcos ends this first chapter with an afterthought pertaining to the response to crisis. In order to maintain the simulacrum of legitimacy an alteration or movement within the image is needed.

> A new figure (which is not a new image, but, the movement of an existing detail to the front) is necessary; the alteration of power which is proposed is, in reality, the alteration of the images in the same mirror, the exchange of the first and second planes in the details of the same image, the same mirror, or power. (2004: 155)

This brief afterword acts as segue to the second chapter on 'legitimate' opposition. Marcos opens the second chapter, the 'Second Mirror', by locating the space of opposition in party politics. Once again positioning crisis, this time the crisis within the Institutional Revolutionary Party (*Partido Revolucionario Institucional* or PRI) – the party that held control of the Mexican state from 1929 until 2000, as opening up a space for oppositional critique and challenge (ibid.: 156). However, while Marcos maintains that the work of the (only true) leftist party, the PRD (Party of the Democratic Revolution), 'opened an important space [for political citizen participation] in Mexico' (ibid.: 157), the party's clamour for legitimization by holding the centre-position trapped it in the same web of neoliberal power (ibid.).

It is here where we see the reason for zapatismo's rejection of representational democracy and party politics. If there are two separate spaces of Mexico: one 'alive' in its service to a business and political people and one 'dead' in its haughty disconnected relationship to the indigenous, rural, working, and poor people, party politics can only be located in the first space. The structure of representational democracies is such that for a party to come to power, it must occupy a centre ground where change is no more than a 'stable alteration' (ibid.), leaving the foundation of neoliberal power intact. Here again, the question of legality comes to the forefront. For a party to have legal recognition, for it to be registered, it must operate within the political centre. This entails abandoning notions of radical change, 'The legal left erases itself and tried to fight to conquer a space which everyone is grabbing' (ibid.).

What remains of the opposition is the 'illegal left', that is to say the leftist groups not officially recognized by the state. The illegality of the left comes from their refusal to accept the Mexican state's 'gift' of law. As Weber notes, the 'gift of law is a gift only if it is accepted as such. Given the extra-territorial situation of the lawgiver, such an acceptance must be brought about through … "persuasion"' (1992: 250). Furthermore, Weber goes on to state that to 'accept the gift of law means that "each individual promises to obey without reservation," *but only under the condition that 'it is for the welfare of all*' (ibid., emphasis added). This means that the people are ultimately the sole judge of the legitimacy law, but also unable to distinguish between the needs of the many and the needs of the few, this means that the social contract is ultimately a conditional promise whose conditions are inaccessible (ibid.). The lawgivers gain their power through simulacrum by claiming to speak for 'the People' where the noun is so universal that it loses any meaning; it is a simulacrum of identity (ibid.: 251). The 'legality' or 'illegality' of a people then represents which Mexican territory they occupy: the alive and mobile or dead and static space.

Marcos acknowledges the importance of the illegal left's political work, its influence in regional spaces and in 'critical junctures' (2004: 157), but he nevertheless maintains a critical analysis. It is worth quoting his analysis at length:

> Both the legal and illegal left share a cannibalistic mirror, one which engulfs all that is nearby, and nevertheless suffers serious digestive problems: it spits up what it eats. All the left which considers itself worthy of the name is the vanguard. This means there are so many vanguards that no one knows where to walk and no 'contingent' exists to follow these vanguards. 'Political realism' and cynicism are more than common places, they are articles of prime necessity. The new left professes the old left and the marks are the tiny mirrors of the great mirror of the opposition in Mexico. (2004: 157–8)

Painting a bleak picture, Marcos concludes by noting that the illegal left, fragmented and confronting itself, has only the honour of not surrendering and of continuing to resist despite the odds and despite itself (ibid.). The illegal left's regression to historical vanguards[4] has failed to significantly alter the mirror space of Power. This is because, like the Mexican state, the vanguard still instils itself as the lawgiver to speak for a generic people. The social contract has no right to promise anything to the people it seeks to create beyond the promise of a proper name, the gathering of all needs and desires into one sole object, but it does so anyway (Weber 1992: 251–2). By submitting to the law of naming, the same law that works to maintain power *over* a people, the vanguard reproduces the same double articulation, the double production of space as neoliberal governmentality. As such, the illegal left is unable to create new spaces, only slightly alter current spaces of power.

An analysis of the space of 'the people' is the third chapter in this essay. Addressing the 'third mirror' to the people 'without a party', to 'civil society', Marcos once again begins with an examination of how crisis opens up a heterotopic space of analysis, but draws attention to the lack of actionable power this space of critique has (2004: 158). The chapter notes that those living within the dead image of Mexico – the indigenous, rural, and poor – are united by their shared economic oppression, but are unable to form into a movement.

Writing that 'what the economy unites, politics separates' (ibid.: 159), Marcos continues to note the lack of a machine at the centre of the people that would generate a shared communal desire; what Guattari would have termed a desiring machine. The people as such lack any subjectivity, existing as isolated individuals (ibid.).

The essay ends with a short chapter on the fourth mirror. Whereas the preceding chapters go into great detail in the critical deconstruction of Mexico and Mexican politics, the final chapter is more poetic, and composed only of 'instructions' to see the forth mirror. The instructions detail the mirror's utopic promise, but rather than simply positioning the mirror as external to the subject so they can see where they are and where they could be, the instructions end by asking the reader to 'look downward, to the left [where] another image will appear' (ibid.: 160) and the reader has been transported to a Zapatista march on Mexico City. This final chapter points to the mirror at the heart of Zapatista subjectivity, a mediation machine that transmits desire.

Pasamontaña

The enigmatic fourth mirror in 'Power as a Mirror and an Image' invites us to examine the crucial function mirrors play mediating Zapatista subjectivity. Turning briefly back to Kittler, one of the more remarkable moves he made was a rereading of Lacanian psychoanalysis that demonstrates how subjectivity is an effect of machines of mediation (1997: 143–4; see also Wilson 2015: 111). This means that the internal space of the heterotopic mirror, the 'invisible' operation of power through the repetition of statements, always comes first. Humans are created in the wake of mediation. This post-human or, better yet, machinic-human theory of subjectivity brings media theory away from McLuhan's humanism and moves it closer to the machinic philosophy of Deleuze and Guattari. Zapatismo also uses this basic understanding of machinic subjectivity, but while Kittler sees the mediation as trapped within the Lacanian Symbolic, where its function is both internal and invisible, the Zapatistas move the mediating machine into a Thirdspace position by externalizing it in the form of their now iconic *pasamontaña*, their ski mask.

The Zapatistas' use of their *pasamontaña* helps to locate and reflect a multiplicity of resistance that draws on their zapatismo philosophy, engendering communication without a transcendental language system of representation (see Guattari 2009: 158; Querrien 2011: 89–90). In other words, to paraphrase Marcos, *the mask becomes a mirror* that reflects and connects all resistant subjectivities (Marcos 2004: 244). Here we have an uncanny similarity to Deleuze and Guattari's concept of 'becoming-everybody/everything'. To become-everybody is, quoting Deleuze and Guattari, 'to make a world (*faire un monde*). By process of elimination, one is no longer anything more than an abstract line, or a piece in a puzzle that is itself abstract. It is only by conjugating, by continuing with other lines, other pieces, that one makes a world that can overlay the first one, like a transparency' (Deleuze and Guattari 2002: 280). For Deleuze and Guattari, to become-everybody is to become an abstract, de-centred, line of flight and to be imperceptible, indiscernible, and impersonal, what they refer to as the 'three virtues' (ibid.: 280). However, there is a contradiction here. To be imperceivable is to be perceived (ibid.: 281). This is because there is a relation established to a 'given threshold of perception, which is by nature relative and thus plays the role of mediation … and makes forms perceivable to perceiving subjects' (ibid.). Thus their deployment of the mask helps to establish a proximity to others on a similar line of flight (a deterritorialization). Geographical distance no longer matters (this is also immensely aided by their prolific and effective use of the internet).

By universalizing their struggle, and adding to it the struggles of every oppressed person through the use of the iconic image of the *pasamontaña* – something that since the WTO protests in Seattle in 1999 has become part and parcel of the anti-globalization movement and often explicitly references the Zapatistas – they have created a subjectivity that has nothing to do with a unified subject teleologically stemming from a mythic common ancestor. It is by the proximity of desire that the Zapatistas enact a new assemblage politics. All of these singular subjectivities within zapatismo are bound together in solidarity, not by any formal relations but through proximity to one another, to those who are closest to what one is becoming, and through which one becomes. This is the sense in which, for Deleuze and Guattari, becoming is a process of desire (2002: 272), in this case, a democratic and non-oppressive, non-fascist, desire. And it is a desire that 'infects' others, rather than spreading along lines of filiation.

By externalizing the mirror to reflect a multiplicity of resistants, zapatismo has opened a fourth space of mediation. The mirror's internal function (generative of subjectivity), its external function (a utopic vision), and critical function (Thirdspace) are all operative and visibly active within this fourth space. In this way the *pasamontaña* should be conceived as a void. As Deleuze reminds us, void should not be confused with lack (Deleuze 2006: 66–7). A void is fully 'part of the constitution of the field of desire criss-crossed by particles and fluxes' (ibid.: 67). Void in this sense is the constitutive space within desire where singularities are able to form new assemblages. This is because space is not constituted by what fills it, but rather by the consciousness of space. An illustrative example of this can be found in Japanese where the word for void (ma) and the word for architecture (ken) are represented by the same character. So, whereas lack signifies absence, void signifies the consciousness of a space of possibilities. Deleuze and Guattari make this spatial understanding of the void evident in *A Thousand Plateaus* when they note that an absolute deterritorialization can only take place within a void (2002: 200). This is why the Zapatista notion of the *pasamontaña* being a mirror works to mediate without representing. The mirror here is essentially a void, it is negative space, or at least the consciousness of space, that has the potential to be filled with any assortment of different resistant subjectivities, utopic visions, and critical engagements. As such, the *pasamontaña* opens up zapatismo to the world and provides the mediated space for new assemblages to form in constant becomings.

Conclusion: Theory to praxis

Throughout their oeuvre, the Zapatistas have constructed a theory of heterotopias and mediation that far exceeds Foucault's and Kittler's. The principle space they use, the mirror, takes on a powerful significance in the formation of their critical response to crisis as well as their way of mediating a resistant subjectivity. But perhaps the most important aspect of the Zapatistas' use of space is the way in which they are able to move from a theory of heterotopias to a practice of creating other spaces. This practice has played out in two very important ways. The first is with their various international gatherings of resistance, international art biennales 'against neo-liberalism', as well as the Zapatistas' international Little Zapatista School of Freedom. These gatherings have brought people resisting neoliberalism from around the world together under the idea that through daily interactions the activists who participated would learn about each others' struggles, their successes and failures, in order to create a world not exploited by neoliberal economics and governmentality. This process of opening up was first made explicit in a communiqué dated 20 January 1994 which read that revolution will arise from 'a chance for a *democratic space* for resolving the confrontation among diverse political proposals' (EZLN in Marcos 2005: 91, my emphasis). Later the same year the EZLN reaffirmed their commitment to opening up a democratic space in their opening address to the National Democratic Convention (CND).

The stated purpose of these gatherings is to *create a space for democracy* in Mexico (ibid.). They described the creation of this space in the following way:

> [We] call for a sovereign and revolutionary National Democratic Convention from which will come a transitional government and a new national law, a new constitution that will guarantee the legal fulfilment of the people's will.
>
> This sovereign revolutionary convention will be national in that all states of the federation will be represented. It will be plural in that all patriotic sectors will be represented. It will be democratic in that it will make decisions through national *consultas*.[5]
>
> The convention will be presided over, freely and voluntarily, by civilians, prestigious public figures, regardless of their political affiliation, race, religion, sex, or age. (ibid.: 49, emphasis in original)

The final, and arguably most important, way in which the Zapatistas have actualized their spatial theories has been through the deceptively simple act of living as though neoliberalism has ended. That is to say, concretely transforming their territory into a living heterotopic space. From 1994 until 2003 Zapatista territory was formally known as the *Aguascalientes* (the autonomous communities in rebellion); however, in 2003 they announced the 'death' of the *Aguascalientes* and the 'birth' of the *Caracoles* and the Good Government Juntas (GGJ) (Ramírez 2008: 264). *Caracoles*, which literally translates as 'snails', were set up as cultural and political centres that act as the public face of the Zapatistas. The *Caracoles* are self-organized and self-governed political spaces that covered over 40 per cent of Chiapas, involving at least 1,100 communities of between 300 and 400 inhabitants each (Dinerstein 2013: 4). The *Caracoles* are territorial spaces for the operation of the GGJ, the Vigilant Commission of the GGJ, schools, hospitals, and the administration of the Zapatista cooperatives (ibid.: 4–5).

> In the Zapatistas *Caracoles*, decision-making occurs at three distinguishable levels. At the local level, each of the many communities of every town elects its own authorities, i.e., the communal agent (*Agente Communal*) as well as representatives to the Autonomous Juntas (*Consejos Autónomos*), the decision-making body. All posts are voluntary. At the municipal level, delegates of each village meet in assemblies, which can last for 3 days, to reach consensus about decision involving design and execution of community projects. Representatives to the Good Government Juntas and the permanent representatives to the five Snails are elected. Finally, the state (*estadual*) level comprises five Snails: Oventic, Roberto Barrios, Morelia, La Realidad and La Garrucha. The Snails are also cultural spaces, gathering schools, assembly rooms, sport and rest zones, health centres, and cooperatives. (ibid.: 5)

It is worth noting that the birth of the *Caracoles* and the GGJ also marked the withering of the EZLN from community decision-making. The GGJ work within the *Caracoles* to administrate justice and mediate conflicts as well as to promote and supervise other educational, administrative, health care, and work needs in the communities (ibid.). These autonomous zones are governed by the Zapatista mandate to lead by obeying a system of direct democratic voting that has given rise to an environment where women's rights, LGBT and queer rights, child rights, education, and health care flourish.[6] Through the *Caracoles* and GGJ, the Zapatistas are creating a living heterotopic inversion of the neoliberal state.

Notes

1 On this point it is important to include Foucault's seminal lectures published under the title *The Birth of Biopolitics*, which analysed the historic formation of neoliberalism.

2 The article can be found in *Fortune*'s online achieve at http://archive.fortune.com/magazines/fortune/fortune_archive/1992/12/28/77310/index.html

3 This was first articulated ten days after the fighting began; they called an 'offensive unilateral ceasefire' and became more explicit in their *Second Declaration of the Lacandón Jungle*. For more on their offensive ceasefire, see Muñoz's *The Fire and The Word* and Holloway's *Change the World Without Taking Power*.

4 Zapatismo has a well-documented antithetical relationship to vanguard-led movements. While this critique can be found in several pieces of writing, it is most explicit in Marcos's 2003 response to the Basque separatist movement the Euskadita Askatasuna (ETA), 'I shit on all the revolutionary vanguards of this planet' (2005: 35).

5 *Consultas* can be translated in English as *queries*, but also means *consultations* and *conferences*.

6 See Dinerstein, A. C. (2013) 'The Speed of the Snail: The Zapatistas' Autonomy De Facto and the Mexican state'. Working Paper. Centre for Development Studies, University of Bath, Bath, U.K.; Marcos, Sylvia. 'The Zapatista Women's Revolutionary Law as It Is Lived Today.' Editorial. *OpenDemocracy*. OpenDemocracy, 22 July 2014. Web. 01 August 2014. https://www.opendemocracy.net/sylvia-marcos/zapatista-women%E2%80%99s-revolutionary-law-as-it-is-lived-today.

References

Baudrillard, J. (2006), *Simulacra and Simulation*, Ann Arbor: The University of Michigan Press.

Colebrook, C. (2018), 'In the Beginning All the World was America', in D. Hancock, A. Faramelli and R. White (eds), *Spaces of Crisis and Critique: Heterotopias Beyond Foucault*, 153–167, London and New York: Bloomsbury Philosophy.

Deleuze, G. (2006), *Dialogues II*, London and New York: Continuum International Publishing Group.

Deleuze, G. and Guattari, F. (2002), *A Thousand Plateaus: Capitalism and Schizophrenia*, Minneapolis: University of Minnesota Press.

Dinerstrin, A. C. (2013), *The Speed of the Snail: The Zapatistas' Autonomy De Facto and the Mexican State*. Working Paper. Bath, U.K: Centre for Development Studies, University of Bath.

Foucault, M. (1986), 'Of Other Spaces', *Diacritics*. The Johns Hopkins University Press, 16(1): 22–7.

Foucault, M. (2008). *The Birth of Biopolitics: Lecturers at the Collège de France, 1978-1979*, New York: Palgrave Macmillan.

Kittler, F. (1990), *Discourse Networks 1800/1900*, Stanford: University of Stanford Press.

Kittler, F. (1997), *Literature, Media, Information Systems*, edited by J. Johnston, New York: Routledge.

Lucero, H. (2002). *Díaz, Porfirio. In: L. Stacy, ed., Mexico and the United States*. New York: Marshall Cavendish, 261–263.

Marcos, S. (2001), *Our Word Is Our Weapon.*, edited by J. Ponce de León, New York: Seven Stories Press.

Marcos, S. (2005), *Conversations with Durito: Stories of Zapatistas and Neoliberalism*, New York: Autonomedia.

Marcos, S. (2007), *The Speed of Dreams: Selected Writings 2001-2007*, San Francisco: City Lights.

Nail, T. (2012), *Returning to Revolution: Deleuze, Guattari and Zapatismo*, Edinburgh: Edinburgh University Press.

Ramírez, G. M. (eds) (2008), *The Fire and the Word: A History of the Zapatista Movement*, San Francisco: City Lights Books.

Reid, J. (2018), 'The Living Space of the Image', in D. Hancock, A. Faramelli and R. White (eds), *Spaces of Crisis and Critique: Heterotopias Beyond Foucault*, 39–56, London and New York: Bloomsbury Philosophy.

Weber, S. (1992), 'In the Name of the Law', in G. Carlson, D. Cornell, M. Rosefeld (eds), *Deconstruction and the Possibility of Justice*, New York: Routledge.

Wilson, S. (2015), 'The Computer that Couldn't Stop: Artificial Intelligence and Obsessional Neurosis', in S. Wilson and E. Ikoniadou (eds), *Media after Kittler*, 111–34, London and New York: Taylor and Francis.

7

In the beginning all the world was America

Claire Colebrook

> *Nearly all of the Iliad takes place far from hot baths. Nearly all human life has always taken place far from warm baths*
>
> (WEIL 2003: 46).

There were two moments (at least) in the 2016 Republican National Convention that might have caused alarm for anyone concerned with what appears to be an intensification of racist Manichean rhetoric. The first, playing on presidential candidate Donald Trump's 'make America great again' motif, was proclaimed by someone who probably knew all too much about lost greatness: 'Former television star Scott Baio' appealed to the audience to 'make America America again.'[1] If America is currently *not* America then it has somehow been diluted; to lose greatness, it seems, is to lose Americanness. This sentiment was reinforced by retired Navy Seal Michael Luttrell: 'The world outside our borders is a dark place, a scary place … America is the light.'[2] One response to such rhetoric might be to say that talk of dark and light, of borders beyond which there is nothing but a 'scary space', and of an America that can be America *again* is almost psychotic in its consignment of anything other than its own interiority to darkness. What I want to do, by contrast, is take these two notions quite seriously, and almost literally. To say that one might make America *America* again is to distinguish between the actual physical land mass and some idea of America – a virtual America. This would be akin to Locke's notion that 'in the beginning all the world was America' (Locke 1988: 99). The notion of a proper original space that falls into subjection to a force or control that is not fully immanent marks enlightenment liberalism as much as it does contemporary notions of resisting a foreign darkness and fear that would arrive from without, and would take the form of a loss and a going backwards in time.

The figure of America as the original world without constraint capable of generating light, law, and goodness *only* from itself could easily be dismissed as ideology. And yet, as John Protevi has argued, the notion of ideology – that there is something like an idea, or mindset that is the basis of political relations – is itself 'ideological' or specific to a particular mode of existence that thinks of the intellect as the ground from which social formations and actions emerge (Protevi 2013: 120). To think of America as an idea, *and* as an idea of light and self-making beyond which there is darkness: what would it mean to take this claim non-ideologically or literally? Let's entertain the idea that America was once great (in Donald Trump's sense), and that it is a land of light. Greatness is, to quote Trump himself, a relation to law that is thoroughly one's own, not 'stolen' by politicians or those who have allowed the country to be open to immigrants.[3] I want to suggest that there is something quite right about this notion of greatness, Americanness, and light.

First, there is and always has been a world of light surrounded by a scary space of darkness. The energy costs required to power what has understood itself to be the first world of enlightenment has secured a space of safety and self-control by stealing energy from elsewhere (ranging from the human energy of slavery and immigrant labour to the ongoing harnessing of resources that creates disaster zones elsewhere). America is a land of light and safety that has created a border (if not a wall) beyond which one might find only a dark and scary place, the war zones, and waste zones that have been necessitated by the creation of the land of the free. America was indeed once 'great', controlling the illegal, indentured, and slave labour upon which it was built, illuminating itself with the world's highest energy consumption per capita, creating waste that it exported beyond its borders, allowing some few of its population to become self-propelled individuals.

Second, this would require adjusting Locke's claim that in the beginning all the world was America. The figure or idea of America as a land of futurity, possibility, self-making, and of a law that is entirely one's own was dependent upon what one might refer to (after Cormac McCarthy) as Mexico. In the beginning all the world was Mexico. That is to say, what is figured today as the land of crime, drugs, weapons, lawlessness, violence, and destitution – the 'Mexico' that would need to be walled off to make America great again – is the condition for the possibility of what calls itself America. In this respect, Cormac McCarthy's truly significant 'end of the world' work is not *The Road* (2006) but *The Counselor* (2013), although one might reread *The Road* more fruitfully in the aftermath of *The Counselor.* Written by McCarthy and directed by Ridley Scott, *The Counselor* details just how much can be lost when one falls on the other side of the law, and how precarious the threshold between the order of law and the utter abandonment of life turns out to be. One way to read this narrative of a counsellor who undertakes one shady deal and then loses everything once he falls into the land of the lawless is both as a moral cautionary tale and as an imperialist and racist horror directed towards what lies beyond America's borders. Once one is in Mexico all compassion, empathy, ethics, and decency seem to disappear without trace. Mexico – it might seem – would be a figure or harbinger of a world that is always at the brink of falling apart. (This is confirmed outside McCarthy's corpus by all those cinematic post-apocalyptic wastelands that depict 'our' world as having fallen into colonization and expropriation – such as *Elysium* 2013 or the 2016 television series *Colony.*) By the time McCarthy writes *The Road* it is as though those threatening border zones of disaster have now become all that is left, and the 'light' of love, familialism, and care is carried only as a dream, and one that might be just too dangerous to entertain: 'He said the right dreams for a man in peril were dreams of peril and all else was the call of languor and of death' (2006: 17). In the world of *The Road* as well as in the Mexico of *The Counselor* a sentimentalism regarding light inures one to the volatility of life but also, in creating a hiatus or suspension of the shudder of existence, creates a lure that precludes an authentic confrontation with the utter contingency of existence. McCarthy's mapping of the relation between light and darkness – especially in *The Counselor* – exposes the utterly parasitic relation between law/light and exposed life. One can believe in a walled-off Mexico, in an America of sweet wholesome light and life only by way of the most atrocious violence. In *The Counselor* spaces of security, luxury, and law have been achieved by way of expulsions and executions that cannot see the light of day. McCarthy's Mexico, like the world in *The Road,* is not some external accident that befalls a once-benevolent land of light. There can only be a world of safe plenty if, somewhere else, there is a proximity to the 'extinction of all reality': 'A concept no resignation can encompass. Until annihilation comes. And all grand ideas are seen for what they are' (McCarthy 2013).

To say that in the beginning all the world is Mexico – the space that was walled off to achieve a land of light and law – is both to draw upon and modify Giorgio Agamben's lament regarding the present. The land of law, sovereignty, political relations, citizens, safety, and light is not only poised precariously against utter abandonment. Zones of abandonment and their apparent indifference to all law and order are the conditions for the possibility of law and light. Agamben's lament is that today what had once been an exception to the bounded polity – an exterior where life is abandoned utterly – has now become the norm of political life. The temporality and spatial figure Agamben charts is one in which a border between inside and outside, bounded political life and bare life, human and inhuman, is effected with the divine fiat of sovereignty. Rather than a social contract that triumphs over a state of nature (where life is nasty and brutish), it is nasty, short, and brutish life that is brought into being with the event of sovereign inauguration. Agamben's diagnosis and prognosis of the present is – if one were to read him uncharitably – akin to the notion that outside the land of light the world is a dark scary place. His dreamed-of futurity would be one where the zone of *indifference* or threshold between the bounded polity and abandoned life was not marked out by a sovereign decision but was rendered immanent. Rather than bare life that appears as the formless and undifferentiated space expelled from sovereign order, Agamben provides several suggestions regarding a new beatitude. What if one might approach others without some prior or overarching conception of the worthiness of their humanity, personhood, or political identity? Each encounter would be experienced in its simple 'thusness', not worthy because of some pre-given category of (say) the sacredness of the human or of life, but in its 'whatever being' (Agamben 1993). Or, what if one were to live one's life not according to a Christian doctrine of rules, but to follow the *manner* of Christ – not *what* he did (prescriptively) but the manner of his being? This, for Agamben, was the challenge of the Franciscan order: rather than following some transcendent legislation one would allow one's life to unfold in this world as if one were already blessed, as if this world were not abandoned requiring external redemption (Agamben 2011). The profanation of the world would allow it to appear as simply thus, with 'life' no longer requiring order from without. To move beyond sovereignty and sacredness would be to refuse borders, thresholds, and expulsions. This would *not* leave us with undifferentiated, or abandoned and bare life. Rather, the threshold of indifference itself would be lived: one might look at an other's life as poised in a very fragile relation between the pure immanence of the simple sweetness of life and the style or manner that unfolds from a life that has the *potential* to give itself form but that has no proper form determined in advance:

> If instead we define the *common* (or, as others suggest, the same) as a point of indifference between the proper and the improper – that is, as something that can never be grasped in terms of either expropriation or appropriation but that can be grasped, rather, only as *use* – the essential political problem then becomes: 'How does one use a *common*?' (Agamben 2000: 115; emphasis in original)

Agamben's turn to immanence is ostensibly post-Deleuzian, a form of politics that unfolds *not* from some already given essence *nor* towards some proper form. In this respect it provides an antidote to nostalgic modes of politics oriented to regaining a lost origin, 'making America America again', or to fascist drives towards a proper and destined future that must be realized regardless of means. In this respect Agamben's 'means without ends' motif is at once a nice shorthand for immanent politics *and* a subtle corrective to some of the tendencies in his corpus. Without a proper end *all* we are left with is what we do, and this is what would distinguish a fragile inhuman immanence from enlightenment immanence. This is more than a historical–philosophical distinction and cuts to the heart of the idea of immanence, and the ways in which one deploys motifs of space, history, and 'America'. Even if we set aside the racist implications of making America America again, the notion of a law and polity that is fully ours and not corrupted by external, accidental, foreign, or derived accretions has one side facing fascism and another that might be deterritorialized.[4] This is why Deleuze and Guattari write about the possibility of micro-fascisms and 'cancerous' modes of the *Body without Organs*. Totalitarian power from above is overthrown; there is a seeming absence of authority and foundation, a free and liberated political field, and yet that very line of escape from power can itself become a tyrant. If 'America' for Locke was a figure for a world that had no law, authority, or sovereign power from above, that very same escape from imposed power could form itself as a new power. A certain purity or turn inwards would create an escape that no longer proliferates outwards, but becomes a lord unto itself. To say that in the beginning all the world was America is at one and the same time a refusal of received power, constituted authority, and happy servitude, *and* a figuration or naming and securing of that power of refusal:

> What makes fascism dangerous is its molecular or micropolitical power, for it is a mass movement: a cancerous body rather than a totalitarian organism. American film has often depicted these molecular focal points; band, gang, sect, family, town, neighborhood, vehicle fascisms spare no one. Only microfascism provides an answer to the global question: Why does desire desire its own repression, how can it desire its own repression? The masses certainly do not passively submit to power; nor do they 'want' to be repressed, in a kind of masochistic hysteria; nor are they tricked by an ideological lure. Desire is never separable from complex assemblages that necessarily tie into molecular levels, from microformations already shaping postures, attitudes, perceptions, expectations, semiotic systems, etc. Desire is never an undifferentiated instinctual energy, but itself results from a highly developed, engineered setup rich in interactions: a whole supple segmentarity that processes molecular energies and potentially gives desire a fascist determination. Leftist organizations will not be the last to secrete microfascisms. It's too easy to be antifascist on the molar level, and not even see the fascist inside you, the fascist you yourself sustain and nourish and cherish with molecules both personal and collective. (Deleuze and Guattari 1987: 215)

A line of escape is, *essentially,* impure; this is because desire is relational: 'Desire is never separable from complex assemblages that necessarily tie into molecular levels, from microformations already shaping postures, attitudes, perceptions, expectations, semiotic systems, etc.' The overthrow or refusal of authority is always this specific mode of escape. If the style or manner of that line itself becomes an object of investment, desire turns in on itself. Put more concretely, one might say that this is what ties left-wing gestures of radicalism, anarchism, and libertarianism to fascist refusals of anything foreign, derived or accidental. In short, immanence and transcendence do not pan out to a new ethical opposition that will replace good and evil. To think immanently is to be without a strict border between good and evil; it is to see the movement of immanence – of escape, of overthrow, of refusal – as an event that destroys what is constituted as unquestioningly good, but always with the risk of falling back into a moralism that identifies evil. In this respect, as I have already suggested, Giorgio Agamben's criticism of contemporary biopolitical subjection, where we have become nothing more than life to be managed by a sovereign power that is our only recourse to law and order is *both* a destruction of a proper humanism that would subject itself to a normative and violent conception of 'man', *and* a way of thinking 'man' as a being who is nothing more than an open potentiality, freed from determination from without, a life that is nothing other than its experience of being simply 'thus'. *Enlightenment* immanence is precisely that 'freedom from imposed tutelage', that comes from the awareness that there is no 'man' other than a pure coming into being, and (at least in its Kantian form) generates a respect for the human liberated from any concrete norm or end. (One might see all the ways in which this pure springing forth from oneself, without norm, law, or determination, could generate a new fascism of self-determination.) A qualified, corrupted, *animal*, or inhuman immanence would not be a movement of pure self-creation, but would become or *escape* always in a relation to other movements of escape (becoming-animal, whereas there can be no becoming-man).

To say that in the beginning all the world is 'Mexico' might offer a different conception of immanence: there is no, and never was, a land of plenty, freedom, self-generation, and pure creation ex nihilo. All conceptions of the good and right emerge from a past over which one never has full control, and which remains occluded. This is not because one does not have adequate knowledge, but because there is no ground or origin to be seen, just an 'untamed genesis' or anarchy (Derrida 2001: 196, 204). The supposed promised land of light on the other side of the wall is what generates ongoing violence in any supposed land of darkness. There is no way out of this scary space. Accepting the immanence of this land is an acceptance of the risk of desire and the political, a line of escape or refusal is never pure. There is no such thing as utopia in a quite specific sense: there is no such thing as 'no place'. This is also to say that it might yet be quite true that *in the beginning all the world was America*, and *also* that America does not exist because there is no beginning. The space of the polity does not emerge in some decisive sovereign constitutive act, but emerges haphazardly – a kludge rather than a work of art: an ongoing, continually adjusted, cobbled together and authorless composition that does a good enough but never ideal job. Fascism is, as Hannah Arendt pointed out, a conception of political *making* that also implies possible unmaking – a return to a clean slate (which would also be a return to America, making America America again). Action, unlike making, is messy; it does not fabricate according to a model, and because it does not issue in a distinct product it cannot be reversed or erased:

> To have a definite beginning and a definite, predictable end is the mark of fabrication, which through this characteristic alone distinguishes itself from all other human activities. Labor, caught in the cyclical movement of the body's life process, has neither a beginning nor an end. Action, though it may have a definite beginning, never, as we shall see, has a predictable end. This great reliability of work is reflected in that the fabrication process, unlike action, is not irreversible: every thing produced by human hands can be destroyed by them, and no use object is so urgently needed in the life process that its maker cannot survive and afford its destruction. *Homo faber* is indeed a lord and master, not only because he is the master or has set himself up as the master of all nature but because he is master of himself and his doings. This is true neither of the *animal laborans*, which is subject to the necessity of its own life, nor of the man of action, who remains

> in dependence upon his fellow men. Alone with his image of the future product, *homo faber* is free to produce, and again facing alone the work. (Arendt 1958: 277)

This might prompt us to a correction: in the beginning all the world was Mexico. In the beginning is the abandoned space that was produced as the zone from which 'the world' walled itself off. If, today, we are approaching the 'end of the world' what we are really contemplating is the end of *America*: a land of high consumption, high production and a freedom that is always freedom *from* anything other than one's own constitutive will. The condition for that *world* – where 'world' needs to be understood in the phenomenological sense of a horizon of projects and meanings that unfold *for me*, in their singular significance for my life and my time – is the *earth*, a land that does not appear as a future of potentialities. Heidegger famously saw world and earth as co-determining: a world opens up because of all I can do, but this world is only possible because of an earth that *gives itself* in appearing as a potential world:

> Tree and grass, eagle and bull, snake and cricket first enter into their distinctive shapes and thus come to appear as what they are. The Greeks early called this emerging and rising in itself and in all things *phusis*. It clears and illuminates, also, that on which and in which man bases his dwelling. We call this ground the *earth*. What this word says is not to be associated with the idea of a mass of matter deposited somewhere, or with the merely astronomical idea of a planet. Earth is that whence the arising brings back and shelters everything that arises without violation. In the things that arise, earth is present as the sheltering agent. The temple-work, standing there, opens up a world and at the same time sets this world back again on earth, which itself only thus emerges as native ground. (Heidegger 2001: 41)

If Heidegger's 'earth' is 'that whence the arising brings back and shelters everything that arises without violation' then it is nothing more than a proto-world. Phenomenology, like today's figuration of the 'end of the world', can only conceive what is other than the world, as a *possible world* and not the inhuman earth. It can only see what is not America (or Europe) as its own horrific nightmare of a third *world*. For phenomenology (despite all its claims) there is no earth; rather, what is outside the 'world' is either Heidegger's ecological fourfold that is the condition for the emergence of world or worlds. This is what allows Jacques Derrida to say that the death of any other person is the end of the world: 'Death declares each time the end of the world in totality, the end of every possible world, and each time the end of the world as unique totality, therefore irreplaceable and therefore infinite' (Derrida 2003: 9).

The *world* in the phenomenological tradition is a horizon of constituted sense, unfolded from a life and subjectivity that is singular and not substitutable. This earth that is left after the end of the world is, at least as figured by Cormac McCarthy and post-apocalyptic culture more generally, 'Mexico': an abandoned space where 'law' (if there is any) is certainly *not* an expression of the living will of the polity, but a force that is barely more than that which secures its own survival. Post-apocalyptic tales depict us all as bare life. But what this narrative epidemic of the end of our world sustains is not quite a social contract myth, where we fall back into a state of nature, but a colonizing delusion. Our 'becoming-Mexico' is deemed to be improper, a state of loss or abandonment. Yet, these third world abandoned zones are the ongoing condition and effect of a Western fetishization of world.

Jacques Lacan once asked us what it might be like to imagine that woman did not exist. Joan Copjec brilliantly explored this in terms of the fantasmatic frame of sexual difference and ethics (Copjec 2002). There is no moment of plentitude from which we have fallen, and no fulfilment that one might attain were obstacles to fulfilment somehow eliminated. The origin is an effect of an imagined prohibition. In the beginning is the experience of loss or fall *from which* one imagines that there must have been a paradise lost. In short, if what we are living through today is *no longer* Europe or America, or if what we are living through today is the end of the world (the end of Europe, the end of America) *then there must have been an America.* Zizek, too, has explored this logic with regard to immigrants, refugees, foreigners, and sundry others: it is not that we blame our less-than-ideal existence on the other, for without the conception of the other as robbing me of my enjoyment there would be no fulfilment or ('woman') that I can imagine as attainable beyond the prohibitions of the present.

> There is something enigmatically utopian in this impossible demand: as if it is the duty of Europe to realize their dream, a dream which, incidentally, is out of reach to most of Europeans. How many South and East Europeans would also not prefer to live in Norway? One can observe here the paradox of utopia: precisely when people find themselves in poverty, distress, and danger, one would expect that they would be satisfied by a minimum of safety and well-being, the absolute utopia explodes. The hard lesson for the refugees is that 'there is no Norway', even in Norway. They will have to learn to censor their dreams: instead of chasing them in reality, they should focus on changing reality (Zizek 2015).

Zizek's point differs subtly and importantly from Copjec's formulation of an ethics of non-dependence (Lacan 1992: 41). For Copjec, to imagine that there is no woman does not amount to learning to censor dreams and changing reality. Changing reality still accepts this world that would be the plane and ground of our adjustments. The problem is not dreams that exist beyond reality, but the impoverished and unimaginative nature of those dreams. There is (to quote Zizek) no Norway, but bringing Norway or America into being, or changing reality, falls one step short of destroying reality. Let's not make

America great again, or America America again, or even bring America into being. Let us stop imagining the 'end of the world' as a world in which America has become Mexico, for *that* end of the world is a mere change of the reality we have. It still falls within the fetishized conception of worlds: that what there is must unfold from a history of dreams of human futurity. Today it might be better to think the end of the world *not* as the task of imagining that there is no America – no longer a land of plenty, no longer an ongoing, and unbridled hyper-consumption and hyper-production with wastes and labour situated offshore – but imagining there is no Mexico. When what we have known as 'the world' ends it will not be that space that the first world walled off, exploited, pitied, feared, and shored up. Rather, this would be a joyous end of the world, saying an affirmative 'no' to all that has come under the concept of world as a horizon of projects, meanings, purposes, and presupposed humanity. What if what were at the end of the world were *not* that site of *third world* desolation that was the figural and literal condition of our liberal existence, but something unworldly?

Notes

1 http://www.cleveland.com/rnc-2016/index.ssf/2016/07/lets_make_america_america_agai.html

2 https://www.theguardian.com/us-news/2016/jul/19/republican-national-convention-day-one-summary

3 'America will only be great as long as America remains a nation of laws that lives according to the Constitution. No one is above the law. The following steps will return to the American people the safety of their laws, which politicians have stolen from them.' https://www.donaldjtrump.com/positions/immigration-reform

4 Jacques Derrida noted that Walter Benjamin's conception of a divine violence that would sweep away the empty and unthinking inertia of political language shared some of its rhetorical and figural force with intensely anti-Semitic notions of renewing the polity from its radical origin: 'And so Benjamin's argument, which then. Develops into a cntique of the parliamentarism of liberal democracy, is *revolutionary, even marxisant,* but in the two senses of the word "revolutionary", which also includes the

sense "reactionary", that is, the sense of a return to the past of a purer origin. This equivocation is typical enough to have fed many revolutionary discourses on the right and the left, particularly between the two wars. A critique of "degeneracy" as critique of a parliamentarism powerless to control the police violence that substitutes itself for it, is very much a critique of violence on the basis of a "philosophy of history": a putting into archeo-teleological, indeed archeo-eschatological perspective that deciphers the history of *droit* as a decay *(Verfall)* since its origin. The analogy with Schmittian or Heideggerian schemas does not need to be spelled out' (Derrida 1990: 1013).

References

Agamben, Giorgio. (1993), *The Coming Community*/Giorgio Agamben, translated by Michael Hardt, Minneapolis: University of Minnesota Press.

Agamben, Giorgio. (2000), *Means Without End: Notes on Politics*, translated by Vincenzo Binetti and Cesare Cesarino, Minneapolis: University of Minnesota Press.

Agamben, Giorgio. (2011), *The Kingdom and the Glory: For a Theological Genealogy of Economy and Government*, translated by Lorenzo Chiesa (with Matteo Mandarini), Stanford, CA: Stanford University Press.

Arendt, Hannah. (1958), *The Human Condition*, introduction by Margaret Cavendan, 2nd edn, Chicago: University of Chicago Press.

Copjec, Joan. (2002), *Imagine There's No Woman: Ethics and Sublimation*, Cambridge, MA: MIT Press.

Derrida, Jacques. (1990), 'Force of Law: The Mystical Foundation of Authority', translated by Mary Quaintance, *Cardozo Law Review*, 11 (1989–90): 920–1045.

Derrida, Jacques. (2001), *Writing and Difference*, translated by Alan Bass, London: Routledge.

Derrida, Jacques. (2003), *Chaque fois unique, la fin du monde*, Paris: Galilee.

Heidegger, Martin. (2001), *Poetry, Language, Thought,* translated by Albert Hofstadter, New York: Harper Collins.

Lacan, Jacques. (1992), *The Ethics of Psychoanalysis, 1959-60. The Seminar of Jacques Lacan. Book 7,* edited by Jacques Alain Miller, translated by Dennis Porter, London: Routledge.

Lindell, Michael K. (2013), 'Disaster Studies', *Current Sociology*, 61(September): 797–825.

Locke, John. (1988), *Two Treatises of Government (Cambridge Texts in the History of Political Thought)*, 3rd edn, edited by Peter Laslett, Cambridge: Cambridge University Press.

McCarthy, Cormac. (2006), *The Road*, New York: Alfred A. Knopf.

McCarthy, Cormac. (2013), *The Counselor (Movie Tie-in Edition): A Screenplay*, New York: Vintage.

Protevi, John. (2013), *Life, War, Earth: Deleuze and the Sciences*, Minneapolis: University of Minnesota Press.

Weil, Simone. (2003), *The Illiad or the Poem of Force: A Critical Edition*, edited by James P. Holoka, New York: Peter Lang.

Žižek, Slavoj. (2015), 'We Can't Address the EU Refugee Crisis Without Confronting Global Capitalism', http://inthesetimes.com/article/18385/slavoj-zizek-european-refugee-crisis-and-global-capitalism

INDEX

CPSIA information can be obtained
at www.ICGtesting.com
Printed in the USA
LVHW06*0303051018
592503LV00008B/74/P

9 781350 021129